Dearest Bill,

    Thanks for all these years of support. Hope you enjoy the book. It's principles are applicable to many industries in the Global 100.

    Keep in touch.

                      Yours
                      Mohan
                      Nov 99

ACTIVITY-BASED INFORMATION SYSTEMS

WILEY COST MANAGEMENT SERIES

James A. Brimson • *Activity Accounting: An Activity-Based Costing Approach*

Douglas T. Hicks • *Activity-Based Costing: Making It Work for Small and Mid-Sized Companies, Second Edition*

Steve Player and David Keys, editors • *Activity-Based Management: Arthur Andersen's Lessons from the ABM Battlefield, Second Edition*

James A. Brimson and John Antos • *Activity-Based Management for Service Industries, Government Entities, and Nonprofit Organizations*

John Miller • *Activity-Based Management in Daily Operations*

James A. Brimson and John Antos with contributions by Jay Collins • *Driving Value Using Activity-Based Budgeting*

Barry Brinker, editor • *The International Journal of Strategic Cost Management*

ACTIVITY-BASED INFORMATION SYSTEMS

AN EXECUTIVE'S GUIDE TO IMPLEMENTATION

MOHAN NAIR

JOHN WILEY & SONS, INC.
New York • Chichester • Weinheim • Brisbane • Toronto • Singapore

This work is dedicated to my parents, who always believe in me, and to my wife, Charu, who gave me the space to create.

This book is printed on acid-free paper. ∞

Copyright © 1999 by Mohan Nair.
Published by John Wiley & Sons, Inc. All rights reserved.
Published simultaneously in Canada.

No part of this publication may be reproduced, stored in a retrieval system or transmitted in any form or by any means, electronic, mechanical, photocopying, recording, scanning or otherwise, except as permitted under Sections 107 or 108 of the 1976 United States Copyright Act, without either the prior written permission of the Publisher, or authorization through payment of the appropriate per-copy fee to the Copyright Clearance Center, 222 Rosewood Drive, Danvers, MA 01923, (978) 750-8400, fax (978) 750-4744. Requests to the Publisher for permission should be addressed to the Permissions Department, John Wiley & Sons, Inc., 605 Third Avenue, New York, NY 10158-0012, (212) 850-6011, fax (212) 850-6008, E-Mail: PERMREQ @ WILEY.COM.

This publication is designed to provide accurate and authoritative information in regard to the subject matter covered. It is sold with the understanding that the publisher is not engaged in rendering legal, accounting, or other professional services. If legal advice or other expert assistance is required, the services of a competent professional person should be sought.

Library of Congress Cataloging-in-Publication Data:

Nair, Mohan.
    Activity-based information systems : an executive's guide to implementation / Mohan Nair.
       p. cm. — (Wiley cost management series)
    Includes bibliographical references.
    ISBN 0-471-32431-0 (cloth /CD-ROM)
     1. Management information systems. 2. Activity-based costing.
I. Title. II. Series.
T58.6.N34 1999
658.4′038—dc21   99-17670
                                                          CIP

Printed in the United States of America.

10 9 8 7 6 5 4 3 2 1

FOREWORD

For the last 20 years at the J. L. Kellogg Graduate School of Management, I have had the pleasure of teaching strategic cost management to several hundred eager and enthusiastic professionals. On several occasions, Mohan Nair and I have shared the stage at Kellogg.

I have noted that the learning about activity-based costing/management (ABC/M) can be best brought about using all dimensions of the process—from concept to implementation challenges—and including people, process, and technology synergies, leading to knowledge and intelligence.

Many executives in the Global 100 are seeking a unifying introductory yet complete understanding of the field of activity-based management. This book provides insights backed by experience, not doctrines, into successful implementation strategies. Rather than focusing on pitfalls, Mohan centers on creating the conditions that enable success.

These days management, faced with numerous well-intended management books, brushes them aside as the "flavor of the month." In my opinion, *Activity-Based Information Systems* by Mohan Nair is a jewel in the area of modern management where measurement, monitoring, management, and maximization for profitability is the paramount principle in these days of intense competition, product proliferation, and technological growth.

ABC/M is the modern mantra for a journey toward maximum profitability for the commercial sector and one that provides strategic and operational cost guidance to the public sector. For the commercial sector, there are two ways to maximize profit: focus on the top line, using revenue drivers, or on the middle line, using cost drivers, or both.

Under the ABC/M context, companies that must manage should focus on managing activities. To do this, they must be skilled in measuring activities. Furthermore, to measure, they must define activities. Activities can be defined only if companies understand the work being performed, not just accounting ratios.

With this foundation of activities, organizations can understand and predict their future only if this information is delivered through regular and predictable information infrastructures called activity-based information systems (ABIS).

This book provides a unified, meaningful, and implementable roadmap to success using ABC/M and ABIS. When I read Bob Kaplan and Tom Johnson's book titled *Relevance Lost,* I found myself wondering when the relevance will be gained.

After reading this book on ABIS, I realize not only is it relevant, it is implementable and useable. Mohan introduces a unique treatment of ABC/M, introducing the concepts of "data richness" and "information poverty," which he calls "data obesity" and "information starvation." The real core of the book lies in the seven steps to successful ABC/M realization. In these critical steps, Mohan systematically synthesizes the full journey toward competent ABIS implementation.

Even though this book is focused on ABC/M and ABIS, it offers insight into a variety of other management concepts. Whether readers are performing business process reengineering/improvement, time-based management, total quality management, yield management, balanced scorecard, thru-put maximization, budget justification, resource management, profitability management, or strategic cost management, the core foundation to all of these initiatives is activity-based measurement.

This measurement methodology is robust, credible to operational teams, and meaningful. This book provides an ABC of activity-based information systems useful to academics, business managers, and consultants. It will be worth reading and rereading.

BALA BALACHANDRAN
Distinguished Professor of Accounting and Information Systems
and Decision Sciences
Director of the Accounting Research Center
J. L. Kellogg Graduate School of Management

PREFACE

Since 1993 I have had the pleasure of teaching in the field of activity-based information systems. Along with Professor Bala Balachandran at the J. L. Kellogg School of Management's executive program, I have responded to countless questions in the field of implementing activity-based costing from executives in the Global 100. From this vantage point of education, I have had to formalize my experiences as a software provider in a growing and transforming industry. This book is a collection of some of the learnings, questions, and answers.

The study of activity-based information systems is unique to the computer industry as it incorporates:

- In-depth knowledge of costing
- An understanding of software
- An appreciation for enterprise use of systems

As the president of a leading provider of software in the world, I have had the pleasure of overseeing over 16,000 software licenses in customers' hands and over 20,000 individuals trained in the use of activity-based technology. I have found that there is no cookbook for a successful implementation but there can be a guidebook for creating the best conditions whereby success has the best chance to occur. After five years of investigation, this book is the result.

WHO SHOULD READ THIS BOOK?

If you are directed or are directing others to take on an activity-based costing/management (ABC/M) program, this book will assist you in the process. In any activity-based costing (ABC) project, there are four types of participants. They are:

1. The economic user—the one who makes the financial purchase decision
2. The operational user—the one in an operating role who drives products or services
3. The technical user—the one who is an information system (IS) professional who has to maintain or sustain the system

4. The sponsor—the one who brings ABC/M into the organization and is its champion

The economic user is usually the one who sets the vision and signs the checks—namely the chief financial officer/chief executive officer/chief operating officer.

The operational user is usually the operational team that needs the information. Much of the technology implementation benefits directly affect this user.

The technical user is usually an information systems manager, chief information officer, or modeler who puts together the image of the enterprise.

The sponsor or coach is usually in the finance organization or a project leader who believes in ABC/M.

This book is focused on the managers (IS managers, chief financial or operating officers, and project leads) who are fighting to increase their understanding of implementing activity-based information systems. ABC/M is not truly useful until it is used continuously and repeatedly by operations teams, maintained by IS teams, and endorsed and championed by the finance department.

Currently very little professional literature can help these managers get a handle on what to look for in implementing activity-based information systems. This book focuses on how to implement activity-based technology and what to guard against.

ABC/M endeavors begin as pilot projects born from the desk of the finance groups within organizations. They gradually move onto massive rollouts that sometimes even reach global proportions. Three areas are vital to implementing and managing ABC/M: people, process, and technology. Technology, the true enabler of implementations, seems to be the most misunderstood and the least documented.

Currently, there exists a good deal of information on ABC. However, careful study shows that most focuses on the people and project issues of implementation. (See the reference section for learning resources.) Yet a careful blending of the technological aspects of implementation with regard to the people and process issues is overdue. Technology is the primary ingredient to moving from a pilot program to an ongoing and sustainable technology implementation.

Today technology has finally caught up to the growing demands of the methodology and the marriage of technology with people and process is beginning. Adding to the confusion, technological complexity is doubling every two years. With activity-based costing hitting the mainstream business models of the Global 500 and beyond, activity-based information systems are now being

PREFACE

brought into focus. With that, CEOs, CFOs, and COOs must learn about the technical and business personalities of these systems and their implementations. Otherwise, this seemingly small and insignificant part of the information system implementation and transition could bring down giants.

WHAT IS IN THIS BOOK?

This book is a result of over five years of investigation and learning. It brings together, in one text, a treatment of activity-based information systems unique to the literature. It is meant to be an introductory text for those who have only a rudimentary understanding of activity-based information systems.

Chapter 1 describes the common terms used in the ABC/M business and speaks to the transformation this industry is undergoing in the next millenium.

Chapter 2 addresses the challenge of information today. ABC has been restricted to financial information views, but the power of correctly designed ABC/M is that it can be an effective way of viewing relevant, operationally potent data. This chapter isolates the real reason why ABC/M is valuable to the challenge of digesting all the various data streams that come at managers. In a nutshell, managers are not suffering from a lack of information; they are drowning in it.

Chapter 3 is a new look at an old subject, the concepts and value of ABC/M. It explains what ABC is and what it has evolved into through the years.

Armed with the knowledge of what ABC/M is, Chapter 4 identifies the fact that implementers go through a life cycle of learning and implementation, graduating from one phase to another. Any violation of these phases tends to lead to failure or a banishment to prior phases to relearn. Chapters 2 through 4 touch on the evolution of these systems and focus on a framework for evaluating and understanding them.

What is an activity-based information system (ABIS)? Systems are usually viewed as computer boxes with keyboards and software. Actually, however, systems are virtual objects that combine all the physical powers of computers, software and hardware, into an architecture that accepts and processes data and delivers information. Chapter 5 describes an ABIS. Chapter 6 includes an ABIS in an activity-based business intelligence system that delivers true ABC/M onto the desktop.

Chapters 7 to 14 share the sevenfold way to design and implement an ABIS. Note that these ways are not necessarily considered sequential. In fact, they can be organized in a reasonably random order. The message behind these seven ways is to "know before you show" and to prepare for the preconditions of success; then it will come. Many times ABC/M projects are high-risk endeavors; if success is not planned for, they will fail 50 percent of the time. The seven ways are seven characteristics of successful projects.

Chapter 15 illustrates four examples of successful ABC/M projects in four different organizations, namely, Willard Foods from South Africa, Providence Portland Medical Center and US Airways from the United States, and Grupo Casa Autrey from Mexico.

The back matter includes reference materials and a checklist that is a kickstart to an ABIS evaluation.

FOCUS ON ACTIVITY-BASED INFORMATION SYSTEMS

Activity-based technology has reached a point where true introspective intelligence about a business can be obtained and maintained. In the world of ABC, computer systems did serve ABC endeavors in the past, but a new wave of technological development has integrated activity-based information systems with data warehouses and desktop business intelligence providing solutions once unachievable. This book:

- Records and documents the evolution of ABIS (activity-based information systems)
- Identifies and defines Activity-based Business Intelligence with respect to data warehousing, business modeling and desktop navigation tools
- Provides examples of implementation of activity-based systems
- Defines a project cycle for ABC/M technology implementations
- Assists the reader in identifying true Activity-based Information Systems built and designed for sustainable versus pilot implementations
- Provides an informal checklist to understand and measure ABC/M vendor offerings

Most of all, this book attempts to bring together the knowledge and focus on activity-based technology and process that are essential to a successful ABC/M implementation.

Regards,

MOHAN NAIR
mnair@msn.com
mohann@abctech.com

ACKNOWLEDGMENTS

I never really considered that the field of cost management could be interesting until I sat in an audience listening to an energetic and enthusiastic presentation to software users. At that time I gained my first understanding of what I now consider the best-kept secret in business—activity-based cost/management (ABC/M).

Professor Bala Balachandran, Distinguished Professor of Accounting and Information Systems and Decision Sciences at the J. L. Kellogg School of Management, was the inspiring speaker who awakened my interest in ABC/M; I also am indebted to Chris Pieper, CEO of ABC Technologies, who was there to educate me about the field whenever I needed help; Steve Player, from Arthur Andersen & Co., LLP, whose commitment and contribution to the industry is constant; Ashok Vadgama, whose work in Sematech, a semiconductor consortium, and in various high-technology semiconductor firms, has fundamentally changed the way the industry views organizational effectiveness; Paul Bierbusse, from Ernst & Young, LLP, who always showed his commitment; and Tom Freeman, who supported my efforts and kindly secured permission from CAM-I for me to use The CAM-I *Glossary of Activity-Based Management* in this book.

When I was first asked to write this book by Sheck Cho of John Wiley & Sons, I imagined that it would take hard work and time. Sheck made it easier. I am grateful to him for his kind supervision and his personal touch. I also want to thank Joyce Ting of John Wiley & Sons for her work on the production end.

Although creating a book was a strain on my family, all I received was unconditional encouragement and support from my wife, Charu, my loving father and mother, and my brother. Charu, your dedication to education fuels me. Papa and Mummy, I write this book as a symbol of your trust and sacrifice in me and my mission. Raj, my brother, thanks for teaching me how to create.

Acknowledgments also go to Professor Michael Porter at Harvard University, Professor Philip Kotler at Kellogg Graduate School of Business Management, and the employees of ABC Technologies worldwide.

The case studies were selected because they were well written by distinguished practitioners. Acknowledgments go to Joe Donnelly (Arthur Andersen LLP) and Dave Buchanan (US Airways), Don Miller (Providence Portland Medical Center), Randall Benson (Benson Consulting), Rick Sahli (Providence Health System), Jorge Medina More E., Ruben G. Camiro (Grupo Casa Autrey), Keith Phillips (National Brands Ltd.), and Kevin Dilton-Hill (World-Class International).

There is no greater teacher than experience. In a way, I gained experience witnessing customers undergo countless projects. This book is not just the result of my experience but the output of the many thousands who took the risks when the tides were against them to prove that ABC/M could be part of successful improvement campaigns. This book is dedicated to these methodology and technology pioneers who are still marching to the "bits and bytes" of their own drums.

Finally, there were many nights dedicated to writing and typing. I am always reminded of the virtues of sacrifice and quiet commitment by my loyal friend and teacher—my dog Kuki. I write of successes while she is one.

CONTENTS

About the Author xvii
1 Introduction 1
2 Challenge with Information 4
  New Frontier of Competitiveness 5
  We Need to Listen to "Moore" 6
  Information Is No Longer Power 7
  What Brings Relevance to Information? 7
  Relevance Is Subjective 8
  Role of Activity-Based Information in Relevance 9
  Operational Value of Activity-Based Information 10
  Users Are Performing ABC But Sometimes They
    Do Not Measure It 12
3 Evolution of Activity-Based Cost Systems 14
  Clip-by-Clip Example 15
  Did Manufacturing Companies Miss the Evolution? 17
  Then Came Activity-Based Costing 18
  The CAM-I Cross—The Symbol of ABC 19
  Hidden Beauty of the CAM-I Cross 21
  Common Realizations Using ABC and the Cross 22
  Is ABC Just Another Initiative? 24
  Comparison of Fads vs. Embedded Technologies 25
  Food for Thought: ABC in the Food and Grocery Industry 27
  ABC Has Utility 29
  Powerful Side Effects of ABC 30
  ABC/M Is Poised for Growth and Adoption 31
  Where Do We Go from Here? 33
  Understanding the Potential of ABC/M 34
4 Activity-Based Management Learning Life Cycle 36
  Trigger Phase 36
  Education Phase 38
  Pilot Phase 39
  Enterprise Phase 41
5 What Is an Activity-Based Information System? 46
  Data Collection and Input Subsystem 47
  Modeling and Analysis Subsystem 49

Reporting and Deployment Subsystems   52
Predictive and Planning Subsystems   53
Infrastructure Subsystem   54
Virtual Nature of Subsystems   54

6 Ultimate Partnership: Activity-Based Costing and Business Intelligence   56
Three Approaches to ABC/M Systems   56
Integrated Enterprise System View   57
Best of Breed: Analytic Applications   58
The Choice Is to Do Both   59
Analytic ABC Applications Remain Post-ERP Implementation   59
What Is Enterprise-Wide?   60
Components of Business Intelligence Software Systems   60

7 Sevenfold Way: Implementing Activity-Based Information Systems   68

8 First Way: Find the Footprint   74
From Pitfalls to Enablers   76
Preparation for the ABC Journey   79
Understanding Self Demands Careful Observation   86
Information Can Behave Differently   89
How to Approach Educating the Organization   89
Moving from Agreement to Commitment   90

9 Second Way: Understand the Evolution of Activity-Based Information Systems   93
It Is a Jungle Out There   93
In the Beginning, Gazelles Roamed the Landscape   93
Giraffes Increased the View   94
Then Came the Software Tigers   94
Strong and Bold Buffalo Decide to Join the Pack   95
Survival of the Species   95
What Does the Future Hold?   96
Deciding Which Software Vendor to Work With   96
Rules of Engagement in Understanding a Vendor   97

10 Third Way: Knowing the Roadmap for Designing an Activity-Based Information System   103
Politics of an ABC/M Program   104
Phases of an ABC Exercise   104
ABC Project Worm   106
Project Managing the Project Phases   124

11 Fourth Way: Treat the Endeavor as a Project   126
Needs of Users Will Always Increase   126

CONTENTS   xv

    Define a Project Schedule with Deliverables   127
    Identify Overall Project Schedules and Systems Design   127
    Develop a Detailed Set of Deliverables and Assign Owners to Each
       Phase of the Project   130
    Define the Level of Involvement for Each Consultant
       and Vendor   130
    Manage the Activity Dictionary   131
    Establish a Tools Inventory   131
    Product and Project Philosophy   132
12 Fifth Way: Watch the Eight Obstacles   133
    Obstacles to Success   133
    Data-Gathering Time   134
    Educating Users   134
    Management Understanding and Support   140
    System Maintenance and Data Replenishment   141
    "Freeloaders" Resist Change   142
    Searching for Push-Button Solutions   142
    Expecting the Model to Freeze   142
    Hidden Costs   143
13 Sixth Way: Align with Strategy   145
    Understanding Strategy and Activities   146
    Strategy and ABC/M   150
14 Seventh Way: Plan for Enterprise-Wide Expansion   152
    Design the Team for Enterprise Expansion   153
    Expand the Model Control Technology   154
    Design the Enterprise Configuration   155
    Not All Worlds Are Created Equal   156
15 Lessons in Success   158
    Case 1: Willards Foods: Managing Customer Profitability with ABC
       Information   159
    Case 2: Providence Portland Medical Center Gets a $1.6m Shot in the
       Arm   166
    Case 3: US Airways Takes Off with ABC   176
    Case 4: Grupo Casa Autrey's CFO Drives Profitability
       Using ABC   181
Appendix A   Checklist for System Selection   185
Appendix B   Informational Web Sites   193
Suggested Readings   194
CAM-I Glossary of Terms   203
Index   220

ABOUT THE AUTHOR

Mohan Nair is President and Chief Operating Officer of ABC Technologies, a leader in activity-based costing systems in the world. In his capacity, Mohan oversees 16,000 software licenses worldwide serving 6,000 customers. ABC Technologies was highlighted as an INC 500 company in 1997, "1996 Oregon Emerging Company" from the Oregon Enterprise Forum, and for three straight years has been one of the top 100 fastest growing companies in Oregon.

Prior to joining ABC Tech, Mr. Nair was President of Emerge Inc., a management consulting firm specializing in strategy, CEO coaching, and leadership. In the last 27 years, he has held pivotal executive level positions in several high performance companies such as Intel Corporation, Mentor Graphics Corp., and Network Associates.

A sought-after speaker, Mohan has also been profiled and quoted in CNBC-Asia, PM Magazine-TV, *Nations Business Magazine, Industry Week,* and *Forbes Magazine.*

He is Adjunct Professor of Business Management at J. L. Kellogg School of Management at Northwestern University in Chicago and the Chair of the Oregon Council of the American Electronics Association. He holds a Bachelor of Science in Computer Science/Business and a Master of Science in Computer and Information Science from the University of Oregon. He is also an alumni of the Advanced Management College from Stanford University.

1

INTRODUCTION

Activity-based information is fast becoming the essential information that operating managers need. Peter Drucker, in a recent article, declared that activity-based costing (ABC) is essential to operational excellence in the next century.[1] Thus far most books written about activity-based costing/management (ABC/M) have focused on:

- Activity-based management concepts and value
- Implementation process and pitfalls
- Case studies

This book fills a very important gap in the learning that few have addressed—the conceptual and practical aspects of understanding and implementing activity-based information systems (ABIS). Knowledge of and experience in information systems implementation is critical to project success. Yet today only a few people understand the formalisms and rigor required for an ABC/M project.

This book helps to formalize the process of this learning. Designed for executives with little technical education, the book focuses on activity-based information systems with a strong grounding for the practical. But first it is necessary to understand basic terminology.

Many times the term "ABC/M" is used in the text. What does it really mean? ABC is the exercise that determines the cost of activities, processes, and cost objects. ABM defines the art of making decisions using ABC information. In simple terms, ABC is the costing work while ABM is the decision made using the information. Hence, ABC/M is used to remind readers of the necessary result of an ABC exercise, that is, making decisions.

Thus far most ABC/M implementations have taken place in the finance community. Over the last decade, the field of study has grown to include operational, information technology professionals and management.

Almost 50 percent of all activity-based costing endeavors fail to reach their true potential. While common belief held that the failures are due to bad implementations, new thought contends that they may have been doomed by design. A failed ABC/M endeavor is a result of a failure in design in the entire system implementation.

Practitioners, consultants, and software vendors are constantly working together to overcome the challenges facing ABC/M projects, but the list of things that can go wrong is long. In fact, over 30 pitfalls already have been identified.[2] This book focuses on how to succeed with the technical aspects of ABC/M or, at a minimum, how to increase the opportunities derived from an ABC/M exercise. This is done by focusing on the approach and design of an activity-based environment rather than the necessary mouse moves to build a model.

This book is about ensuring a successful information system-based implementation, not by avoiding failure but by designing the project for success. Key variables assist in establishing the consequence of an ABC/M project; when enhanced and controlled, these key variables work to your advantage and increase the probability of success.

Usually ABC/M projects are delegated to visionary finance professionals who leave the caves of finance and enter the open skies of operations only to realize that other clouds hamper communication. Each reader could be one of these champions of change. Through the years, many visionaries have built a storehouse of knowledge and registered successful implementations. Successful implementations move ABC/M out of the finance office to include the operational teams and the information systems (IS) teams, using the common understandings and methods defined in this book. They have found sustainable value from an ABC/M exercise.

Years ago, academic thought leaders supervised ABC endeavors; then came the consultants, who spread this knowledge with countless initial pilot implementations. Practitioners evolved in many of these pioneering companies, sometimes moving from one to another because they were so much in demand.

In the most recent years a transformation in ABC projects is occurring. First, these projects are a global phenomenon. Second, the projects are attempting to take ABC/M to many different plants and offices worldwide. Third, they are not asking about conceptual proof but moving directly into implementation proof. Fourth, a larger portion of endeavors use in-house expertise to implement these endeavors. Fifth, the technology implementations are seeking to be more elaborate and sophisticated. Finally, enterprise resource planning (ERP) companies have entered the market, which has raised the attention of chief financial officers and chief information officers of the Global 100.

All in all, ABC/M has become a vehicle for competitive advantage in almost all industries ranging from government, to manufacturing, to process manufacturing, to service industries. In the past, ABC/M implementors had the luxury of time because everyone was not as knowledgeable on the subject. Now the marketplace winners and losers will be defined by who can implement and derive the value of ABC/M using activity-based information systems.

NOTES

1. Peter Drucker, "The Information Executives Truly Need," *Harvard Business Review* (January-February 1995), pp. 54–62.
2. S. Player, *Activity-Based Management: Lessons from the ABM Battlefield* (New York: Mastermedia Ltd., 1995).

2

CHALLENGE WITH INFORMATION

It is estimated that only 3 to 5 percent of corporate information is analyzed. Why is this not a surprise? Examine any business in the 1990s. Executives are inundated with faxes, electronic mail, telephone messages, conference proceedings, direct mail, telemarketing calls, paper mail, and reports. In fact, if they actually read and analyzed everything they received, they would not do anything productive to improve organizations. Just when they thought they had control over information, the Internet revolutionized information accessibility and is transforming the very way in which business is performed. Now executives find themselves surfing the Net for hours to trap information.

Business is not getting any more manageable. With corporate intranets, extranets, and knowledge network technologies entering the information management landscape in the Global 100, corporations will never die from starvation when it comes to information. They may die from indigestion. Too much data too quickly, with no analytical framework and no action, seem to be the death slogan for the Global 100.

Companies suffering from data obesity and knowledge starvation have the same symptoms:

- Data disintegration—Are there limited and predictable sources of information or must the organization reinvent processes to find information every time a problem is to be solved?
- Content sensitive information—Is information received immediately applicable for developing measures and metrics or is preprocessing required?
- Fitness of source—Is information credible?
- Depth of information—Does the information pose new questions?
- Data dimensionality—Does the information provide dimensional views and perspectives?

- Timeliness—Is the information one quarter too late?
- Data usefulness—Is the information used regularly?

Equally important is information, but it is not always viewed as an asset. Recently the push for more and more information has had some negative effects:

- The value of information diminishes with time. Old, untimely information can be extremely destructive to the natural flow of business. Assumptions are made based on data, and assumptions based on untimely information could halt the successful momentum of a company's actions on products and services.
- Information may have negative value when it is not only untimely but wrong. "Misinformation subtracts value from the valuable."[1] Wrong or outdated information may lead users to the wrong conclusions.
- The value of information is relationship dependent; that is, finite data is useless without the correct context and the relationship of the finite data to other finite data. For example, knowing about costs overruns in a factory is relevant, but more relevant is understanding where and what caused them.

In a nutshell, information unused and unrelated is a depreciating asset and can turn into a liability very quickly.

NEW FRONTIER OF COMPETITIVENESS

In the past, business enjoyed increasing market share and huge profits. With global competitiveness splitting the market pies, these companies are fast realizing that they must do more with what talent and tools they have. In the search for the ultimate "magic pill," whether it is operational efficiency, gaining loyal customers, building a new mousetrap, or establishing a powerful value chain of vendors and suppliers, companies have discovered that the true lasting competitive advantage is not just the abovementioned strategic themes but knowledge. Knowledge has been the theme song of management gurus of the 1990s. But knowing without doing can be a waste of time and energy.

Beyond this discovery, the Global 100 is fast realizing that "self-knowledge" and applied self-knowledge is true power—companies must know themselves better than their competitors know them to act on strengths effectively in the marketplace. For example, Wal-Mart changed the way manufacturers, brokers, retailers, and wholesalers performed work. Wal-Mart changed the entire business model and activities in the $500 billion food industry. Knowing what the company did well and knowing what its competitors did not know about the consumer brought Wal-Mart to victory with a 3 percent profit margin compared to its competitors' with less than 1 percent margin.

Wal-Mart has sent the food industry into a cost-cutting efficiency adventure. Wal-Mart used its self-knowledge and applied it for customer retention. More than information technology, Wal-Mart understands how to get the best from its technology, vendors, and customers better than some others do.

WE NEED TO LISTEN TO "MOORE"

Gordon Moore, cofounder of Intel Corporation, introduced the notion of complexity growth when he declared that the microprocessor would double in complexity every two years.[2] The prediction has been borne out. It is believed that in the years to come, a single desktop computer will be more powerful than all the computer power combined in the world today. Similarly, it is believed conservatively that the amount of private and corporate data stored on computers is doubling every 12 to 18 months. Clearly, it is not a lack of information that holds corporations back.

Nor is it information technology. Faye Borthick, professor of accounting at Georgia State University, and Harold Roth, professor of accounting at the University of Tennessee in Knoxville, declare that "For the first time, information technology is sufficiently well developed that accountants can have the information they want."[3]

With the introduction of data warehousing, data marting, data mining, on-line analytical processing, three-tier client-server technologies, desktop navigation tools, search engines and hardware technologies, information technology seems to have popped up to consume data and expel it to anyone at any time and anywhere. Chapters 4 through 6 will describe these factors further.

These technologies coupled with all the information overload will only bring users irrelevant data faster. Winning companies do not win by mastering quick access to information; they master the ability to, at a sustainable level, provide *relevant* information to the right people at the right time for the right managerial decision.

Peter Drucker, the father of modern management, stated that what is important is not tools; it is the concepts behind them that are important.[4] He declared that today's information lacks a conceptual map to give it relevance to the decision maker. In some ways, the technological treadmill is going faster and faster, almost outstripping the needs of business and creating a life of its own.[5] This new market demand for executives to be powered by information to win gives birth to knowledge leaders, who drive a business using analytical information as guide. Knowledge leaders now must understand that the fundamental competitive capability using these newfound tools is not how much information is gathered but that the leader must optimize the mean-time-between-decisions (MTBD). They must improve how fast they can turn data into decisions to create a new landscape for competitors to chart or be lost in the maze of information.

INFORMATION IS NO LONGER POWER

Today knowledge leaders cannot be measured by what information is obtained and dispensed but by what information is rejected, which will account for significantly more. Without keen selection capability, knowledge leaders will be crushed under the sheer weight and demand of decisions to be made. Consequently, those organizations that master the ability to understand themselves enough to make decisions and command themselves enough to act decisively and consistently will win. Information does not offer power. Decisive actions using relevant information is power. Competitive advantage is best developed in the acquisition and deployment of relevant information to all who need and decide/act with it. Information that used to be held by business analysts will shift dramatically to all managers and decision makers. There is no longer time for hierarchical decision-making protocols. The hierarchy holds the old bones of the corporation in place while the nervous system, where information flows in the company, fights the real wars of wealth acquisition. Relevant data are the fuel for this activity.

WHAT BRINGS RELEVANCE TO INFORMATION?

Peter Drucker contends that information should challenge basic assumptions and have links to strategy. He declares that ABC/M is such information.

Drucker states that enterprises are paid to create wealth, not control costs. But this permise is not reflected in traditional measurements. First-year accounting students are taught that the "balance sheet portrays the liquidation value of the enterprise and provides creditors with worst case information. But enterprises are not normally run to be liquidated."[6]

Drucker seems to believe that information is used for wealth creation. He breaks up information value into four main value categories. They are:

1. Foundation information: diagnostic, cash-flow
2. Productivity information: resource productivity
3. Competence information: measurement of the unique ability that customers pay for
4. Resource-allocation information: managing scarce resources for the current business

Note that he believes these categories to be information on the current business condition and hence tactical in nature. ABC/M practitioners will declare that the greatest impediment to projects is the lack of upper management support. Upper management prefers strategy to costs and hence must see the relationship before supporting a cost-oriented project. The questions surrounding the relationship of ABC/M to strategy are discussed in Chapter 13.

Simply put, today many organizations are running forward but looking backward. These companies are blind to their true product costs, their profitability, and their channel behavior. They cannot answer the most basic questions, such as "Is the cycle time for products and the cost of product creation correlated?" (i.e., do cycle time and product creation cost track with one another? If so, what are the activities that drive these costs and profit? Are they value-added activities or not?)

John Whitney, professor of management at Columbia University, hit the nail on the head when he declared that "I have found that perhaps most businesses do not know the true accrual profit of their products and services, and fewer still know the profitability of customers."[7] ABC/M goes one step further by providing a relationship between these answers and the work performed or the activities that organizations can affect.

In the commercial sector, information is used to uncover these issues; in the public sector, profitability is of no relevance. Budgets take precedence. Questions such as the following arise in the public sector: "How much will it cost to build this road and at what rate per mile?" "How ready are we for a battle?" "Where are our critical activities?" "Can we do more with less federal funding?"

RELEVANCE IS SUBJECTIVE

Essentially, relevance is in the eyes of the beholder. Ultimately, decision makers at the strategic, operational, and financial corners of an organization need information relevant to the decisions they must make. Morris Treadway of PricewaterhouseCoopers LLP describes relevance as "data suitable for a user's need."[8] This may sound anticlimactic, but let us explore this point further. In a nutshell, we know that any relevant information must:

- Link to strategy
- Be linked to activities and groups of activities
- Measure the loss of not doing any particular activity
- Support the four main categories of Drucker's value model
- Feed and support an underlying concept

ROLE OF ACTIVITY-BASED INFORMATION IN RELEVANCE

Treadway declares that "Activity-based Management is both a mind-set and a tool for managing a market-driven business that is grounded in the work being performed."[9] In many ways, ABC/M provides relevant information because it shows "the relationships among activities that are performed, the resources to be managed, the products/services produced and other variables in the business performance mix. Underlying this is that relevance is relationships that reflect my business."[10] We think of business as a flow of resources into the creation of greater resources and profit but we manage these resources as one-dimensional revenue, costs, and profits. Generally, organizations tend not to investigate and understand the true capability of the activities within to achieve their objectives. Furthermore, the basic cost figures serve little value to any operating teams faced with a need to improve the bottom line.

Fortunately and unfortunately, ABC was born on the desk of accountants, who identified it as a weapon of rebellion against the general ledger's weakness for operational expression. However, the rigor of finance gave ABC a strong grounding to enter the Fortune 100 and remain a force for many years. Without this rigorous underpinning, ABC might have gone to the same grave as other initiatives. ABC/M is now being revisited by users with an operational perspective and toolkit. With its new operational and financial identity, it can now become the underpinnings of strategy. (See Chapter 13.)

However, when people think of cost, the first verb that comes to mind is "cut." When people think of activities, the first word that comes to mind seems to be "optimize." ABC, for the last decade, has been guided by cost-focused teams built to "cut." The true value of an ABC/M exercise is to understand activities and their drivers. Finding the knobs to turn on a business will free the locks holding back an organization's performance. In a critical sense, the "A" in ABC is sometimes more important that the "C." Activity management is the language of operational teams. They are measured daily on the results derived by optimizing their activities and processes.

John Whitney affirms this value proposition, saying "In the process of clearing, managers should re-discover those activities that must be not only preserved but also cultivated."[11] He confesses that: "Simply put, companies have no system in place for gathering and processing these data, even though activity-based costing has been around for more than a decade." This is changing, and rapidly. Today systems exist to monitor and record almost anything. In this transaction-oriented society, people love to know everything about everything. Given this preoccupation with data and transaction-based data, any system will crumble. Activity-based systems can carry weight but are designed for relevance because they link activities to products/services and enable resources to be measured by their contribution to the whole. In a sense, "If it does not fit, omit!" should be the motto of an ABC/M program rather than "If it's transaction, it can be used as information."

Transaction systems are designed for bulk retention and storage. Certain systems demand this and deliver what is needed. Analytic applications such as activity-based systems feed the business community's need for relevance and relatedness at a macrolevel.[12] Questions regularly arise with respect to positioning enterprise resource planning (ERP) systems with analytical systems in organizations. Chapter 6 discusses the partnership of ERP systems and analytic applications.

OPERATIONAL VALUE OF ACTIVITY-BASED INFORMATION

The general ledger has been used for centuries. Corporations use the general ledger as a measuring stick comparing the performance between and within companies from one period to another. The statutory bodies measure the performance of companies the same way and have done well through the years in creating a standard way to report and respond to the public. One aspect of the general ledger that has failed to fit modern businesses' needs has been in operational teams trying to understand their true personality and relating this personality to the organization's productivity—its activities, products, and services. Consider Exhibit 2.1 in describing the corporation's financial health. As a stockholder, this information may be all that is relevant. But the operational vice president of the company has been asked to reduce the expenses by 20 percent in order to improve profitability. What actions should be taken?

Many would go for the biggest dollar item to shave dollars—namely salaries. This means headcount, which is a euphemism for human beings with families and pension plans. Corporations do this all the time because, beyond all other rules, Professor Bala Balachandran declares that there is one golden rule: "If the bottom line is negative, the middle line must have been greater than the top line." More information is provided in Exhibit 2.2, which supplies more interesting insights into the nature of the general ledger. Now there are verb noun categories to review regarding work that is being performed by the

HOW DANA DISCOVERS WHAT ITS TRUE COSTS ARE
The material-control department in Plymouth, Minnesota

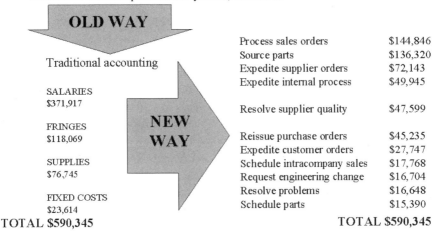

Source: Terence P. Pare, "A New Tool for Managing Costs," *Fortune Magazine*, June 14, 1993.

Exhibit 2.1 ABC vs. Traditional Accounting

resources provided from the general ledger. You may ask, "let's take a look at _____" more deeply. Now the vice president is armed with more information regarding the work itself. Users with more relevant information can ask more questions before taking action. Any ABC/M exercise leads users to ask the next question and the next question until they reach an answer, as compared to the general ledger that leads to a unidimensional action—cut by reducing a number or two. The exploration may uncover activities that can be improved and/or removed. Here the focus is not firing people; it is optimizing activities, so in a sense activities are being fired. Chris Meyer, author of *Fast Cycle Time* and *Blur,* emphasizes that "Marketing tracks market share, operations watches inventory, finance monitors costs and so on. Such results measures tell an organization where it stands in its effort to achieve goals but not how it got there or, even more important, what it should do differently."[13]

Furthermore, what is shown on the left side of Exhibit 2.1 gives no clue as to the nature of the business, i.e. do they bake cakes, run pipes, or sell bonds. The right-hand side of Exhibit 2.1 explains more what work the company is performing and to what level of efficiency. Clearly, activity-based information can provide this type of opportunity. Note: Activity-based costing exercises do not always result in cuts in activities only; like any measurement exercise, sometimes ABC/M results in firings and layoffs. But ABC/M significantly increases probability of not firing/laying off the wrong team, and many times it leads to redirecting resources with a successful bottom-line impact.

1. GENERAL LEDGER AND LEGACY SYSTEMS SERVE A USEFUL PURPOSE: STATUTORY REPORTING AND TRANSACTION PROCESSING.

2. ABIS SERVES STRATEGIC AND OPERATIONAL USES.

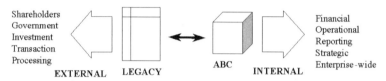

Exhibit 2.2 Balancing General Ledger and ABC

Exhibit 2.1 refers to financial reporting; Exhibit 2.2 refers to the power of ABC/M for managerial, operational reporting. Both are important, both are connected, and both serve different masters.

USERS ARE PERFORMING ABC BUT SOMETIMES THEY DO NOT MEASURE IT

Public companies serve two masters as well: (1) internal and (2) external. The external masters are the statutory bodies, namely the Internal Revenue Service in the United States and the shareholders who expect their companies to report on their status in GAAP terms. Shareholders have been trained to regard such information important as well. But people need more than that to understand a company. Take a look at an annual report. The typical annual report contains two segments:

1. A detailed description of what business this company is in at the front of the document
2. A description of the financials at the back

These beautifully designed documents, with quotes and pictures and drawings of products/services and people, are a mainstay because without them we cannot tell what this company does. Reading the financials leaves people with a taste for the fundamental activities of the business.

Despite all the sophisticated systems available, chief financial officers (CFOs) lack the most logical and basic information. A survey developed by Computer Sciences Corporation and the Financial Executives Institute uncovered that 57 percent of 400 CFOs surveyed declared that their biggest constraint is the inability to establish product and customer profitability. The study also showed that 41.5 percent wanted effective ABC/M systems.[14]

Clearly, GAAP statutory reporting and ABC provide financial information combined with a business description. In a sense, Exhibit 2.2 is an annual report, only in reverse—it takes the financials and provides an activity-based view of a business. Both are related and tied together; one measures and declares the financial position of an organization to external shareholders while the other links it to the work being performed.

NOTES

1. Rich Willis, "Major Boo-Boo," *Forbes ASAP,* April 7, 1997, p. 36.
2. Gary H. Anthes, "The Long Arm of Moore's Law," *ComputerWorld,* October 5, 1998, p. 69. Moore's Law. Moore identified this theory, which is used extensively to identify the growth of semiconductor complexity.
3. Fay A. Borthick and Harold Roth, "Faster Access to More Information for Better Decisions," *Journal of Cost Management* (Winter 1997), p. 25.
4. Peter Drucker presented this notion in his keynote speech at the Annual Users Group meeting for Cognos Corporation in 1997.
5. Lawrence S. Lyons, "Creating Tomorrow's Organization: Unlocking the Benefits of Future Work," *Leader to Leader* (Summer 1997), pp. 7–9. "A gap existed between the needs of the business and the capabilities of technology. Today all that has changed. The capabilities of information technology now outstrip the needs of business."
6. Peter Drucker, "The Information Executives Truly Need," *Harvard Business Review* (January-February 1995), pp. 54–62.
7. John Whitney, "Strategic Renewal for Business Units," *Harvard Business Review* (July-August 1996), p. 85.
8. Morris Treadway, *A Primer on Activity-Based Management: ABM in Utilities: A Process for Managing a Market Driver Business* (Coopers & Lybrand, 1995).
9. Id.
10. Brad Ackright, "KCPL Restructures G.L. to Reflect Activities," *As Easy as ABC: ABC Technologies Newsletter* (Fall 1996).
11. Whitney, "Strategic Renewal for Business Units."
12. Henry Morris, "Applications and Information Access: Information Access Tools," *IDC Bulletin,* No. 14064 (August 1997).
13. Christopher Meyer, "How the Right Measures Help Teams Excel," *Harvard Business Review* (May-June 1994), p. 95.
14. The First Annual Survey of Technology Issues for Financial Executives 1998, Financial Executives Institute, Morristown, NJ and CSC, El Segundo, CA.

3

EVOLUTION OF ACTIVITY-BASED COST SYSTEMS

Activity-based costing (ABC) was developed as a practical solution to managing overhead. In the 1980s, many companies, based on the findings of Robert Kaplan of Harvard Business School, Robin Cooper, Claremont Graduate School, and Tom Johnson of Portland State University, began to realize that traditional accounting systems and cost management methodologies were distorting how overhead should be associated with the product and services the company performed.[1] This is not due to incorrectness but because the nature of overhead had changed while the methods that treated overhead had not. Traditional systems did not evolve to support the changing behavior of costs. In the past, managers had to put up with "overhead" charged to their departments while they knew well that these costs were incorrectly allocated to them.

In the 1980s, the Consortium of Advanced Manufacturing-International (CAM-I) defined ABC as "a methodology that measures the costs and performance of activities, resources and cost objects."[2]

Spurred by articles and books and a great need in the field for an answer to where overhead is going, ABC began to be viewed as an initiative in the 1990s.[3] Unfortunately, billed as a replacement for then-current cost management methods, ABC began to take on the general ledger. This did not work. Even though the industry has moved beyond this, new discoverers of ABC still ask "Does it replace the general ledger?"

Beginning in the manufacturing industry, ABC served a strong need for firms that were struggling to identify a means of:

- Measuring how products and services consume overhead
- Understanding the true costs of activities within organizations
- Understanding the true costs of products and services

- Understanding the true profitability of channels, products, and services
- Quantifying, measuring, analyzing, and improving business processes

The early 1990s were filled with ABC endeavors that were billed as change initiatives that would reengineer the finance output. These initiatives moved from the pure manufacturing companies to cover the process manufacturing industry, the service industry, and the public sector. They were generated by visionary finance teams and champions as a cost-cutting initiative and endorsed by chief financial officers, who were challenged to improve the profitability of their corporations, or in government, to justify budgets. ABC would be used to identify dreaded overhead and assign this large and undefined beast into its correct cage. ABC served a strong purpose then since traditional cost methodologies tended to allocate costs directly to products and services with a single-stage allocation. Costs are allocated based on labor or standard overhead volume drivers. Labor hours, traditionally, being the larger portion of total overhead mix, would drive the decision of where to put overhead costs.

A historic description of the evolution of ABC is found in *Implementing Activity-based Cost Systems* by John Miller and in *A Guide to Total Cost Management* by M. R. Ostrenga, Terrence Ozan, Robert Mcilhartan, and Marcus Harwood.[4]

CLIP-BY-CLIP EXAMPLE

A simple example can illustrate the impact of overhead on costs. Exhibit 3.1 illustrates the challenges of a manufacturing firm that produces paper clips. This firm, NAP (Nair's Papers), builds two kinds of clips: (1) a standard stainless steel clip and (2) a hot pink plastic-coated clip.

Exhibit 3.2 shows a typical analysis based on volume drivers. With a total overhead resource of $1,000,000, the higher-volume product, the stainless steel clips, take most of the costs. This would total $950,000; the cost of the specialty product is assumed to be much less.

Unfortunately the stainless steel clips are not as profitable as desired while the specialty product is doing very well. Exhibit 3.2 shows that the firm is generating a profit of 2.5 cents per clip.

The mainstay, cash-cow stainless steel paper clip product manager has a midlife crisis and has nightmares of being transported to a fourth-world country whose labor costs are low. When the board meets, its members may decide that NAP should deemphasize the stainless steel clips and start to focus on specialty products as their profit margins are worth the work.

EVOLUTION OF ACTIVITY-BASED COST SYSTEMS

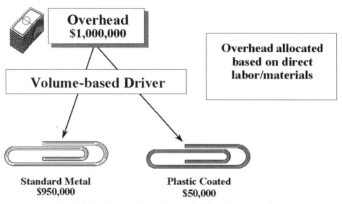

Exhibit 3.1   Traditional Costing Model

Traditional costs management systems are not wrong, just outdated. They have not kept up with the change in assumptions. Because of this, they:

- Undercost complex products and subsidize low-volume products
- Do not reveal hidden costs
- Misdirect performance by emphasizing labor
- Focus on costs, not the activities that cause cost

After a layoff, often people say, "Harry's gone but the work still remains." This is an oversimplified way of stating a very critical mind-set—that labor is no longer the major contributor to overhead—what the labor *does* is the contributor. Hence, firing Harry did not remove the work. Since the activities cause costs, the costs remain.

	Standard Metal	Plastic Coated
Burdened labor	$950,000	$50,000
Units produced	10,000,000	500,000
Unit cost	9.5 cents	10.0 cents
Retail list	10.0 cents	12.5 cents
Profit margin	.5 cents	2.5 cents

Message
 • Specialty clips generate 5X profit of standard clips
Action Plans
 • Deemphasize standard metal clips
 • Refocus production toward plastic clips

Exhibit 3.2   Typical Analysis

DID MANUFACTURING COMPANIES MISS THE EVOLUTION?

In the past, direct labor was a large percentage of the overhead mix. Labor drove business overhead of products. So, where people worked was where the costs were. Hence, costs were allocated using labor as the driver. Other overhead items, contributing only 20 percent of the mix, were allocated in the same distribution as labor.

As the years progressed, manufacturing firms automated work. With this, labor was allocated to other functions besides work itself: to functions like machine setup, preparation, and relationships to vendors. The percentage they contributed to overhead fell when measured against other areas where overhead grew. Soon they became only 20 percent of the overhead mix with other overhead driving 80 percent of the mix. Exhibit 3.3a and b show this transformation. Thus, using traditional labor to drive actual product costs uses only 20

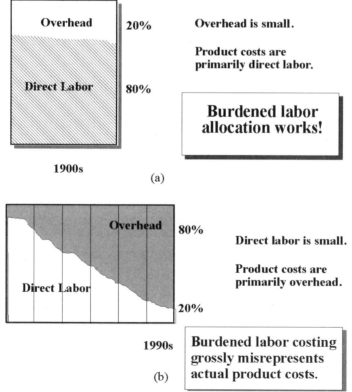

Exhibit 3.3 (a) Early View of Manufacturing Overhead and (b) Evolution of Manufacturing Technology

percent of the total overhead to drive all of overhead. This grossly misrepresents the true costs of products and services.

THEN CAME ACTIVITY-BASED COSTING

Consider the following illustration. The chairperson of NAP instructed the company to take a strong look at ABC. He hired a consultant from the consulting firm Knowalot LLP to take a strong look at the firm's activities. Knowalot found three main activity buckets: (1) special handling, (2) rework, and (3) plastic coating, activities applied only to the plastic-coated clips. After six months of interviewing teams, a host of presentations, and many lunch meetings, Knowalot presents the slides shown in Exhibits 3.4 and 3.5. To everyone's surprise and amazement, using ABC—that is, going to a three-stage model using resources, activities, and products—shows that certain activities consume certain resources but, more important, that certain products consume certain activities using certain other unique or shared drivers. For example, the consultants discovered that plastic-coating activity is not used by stainless steel clips and yet that department was charged for plastic coating in the standard allocation of overhead costs. With ABC, these costs were assigned to plastic-coated clips through the activity of plastic coating.

At the next board meeting, the chairperson claims that, based on a renewed sense of detail, NAP can halt all extensions of plastic-coated clips until further investigation. NAP has no room for unprofitable clips. At a loss of 17.5 cents per clip, he exclaims that the company should just mail the customer a check and stop production. The board discovers that diversity does cost.

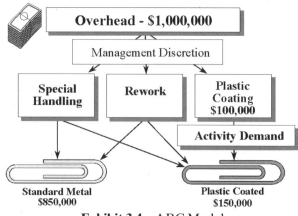

Exhibit 3.4 ABC Model

THE CAM-I CROSS—THE SYMBOL OF ABC

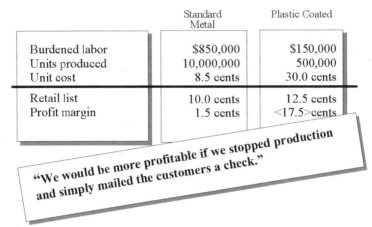

Exhibit 3.5 ABC Analysis

The standard-clip product manager stops writing his resume and increases his production goals while the plastic-coated clip product manager is asked to optimize his business model to profitability.

NAP Inc. learned that the old cost management methods needed to evolve to ABC because a simple single-stage allocation based on volume drivers would not provide the necessary detail for operational effectiveness. With ABC, using a three-stage assignment method, NAP would get a view of how its resources were consumed by its activities and how its activities were consumed by its products and services.

THE CAM-I CROSS—THE SYMBOL OF ABC[5]

The CAM-I model serves as the essential backbone of any activity-based system.[6] Exhibit 3.6 illustrates its components. Often called the CAM-I cross, it can be broken into two main views: (1) the cost assignment view and (2) the process view.

The simplicity of these views hide their profound value to any ABC endeavor. In the paper clip example, these three stages were identified in the cost assignment view. Let us define the parts of the cross in the assignment view:

Resources   Economic element that is applied or used in the performance of activities: salaries, technologies, facilities, supplies.

EVOLUTION OF ACTIVITY-BASED COST SYSTEMS

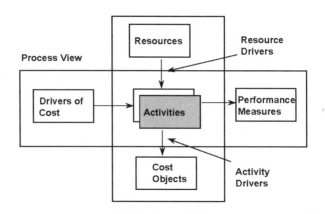

Exhibit 3.6 CAM-I Cross

Activity — Work performed within an organization: unloading trucks, training employees.
Cost objects — Any customer, product, service, project, or work unit.
Resource driver — A measure of quantity of resources consumed by an activity. Sometimes percent time is used as this driver. (See Exhibit 3.7.)
Activity driver — A measure of the frequency and intensity of the demands placed on activities by cost objects. (See Exhibit 3.8.)

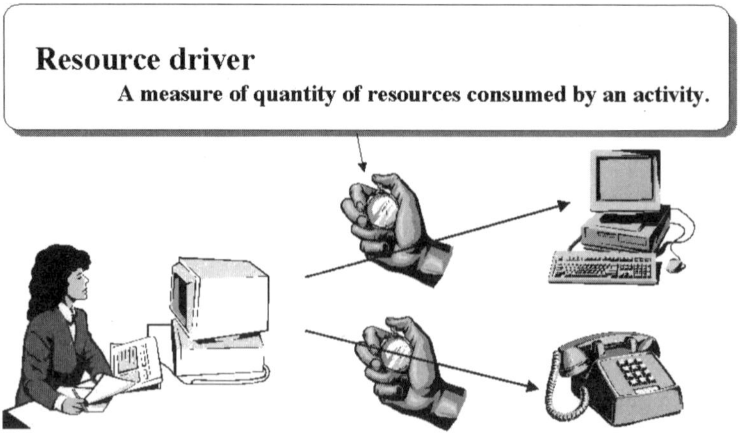

Resource driver
A measure of quantity of resources consumed by an activity.

Exhibit 3.7 Definitions

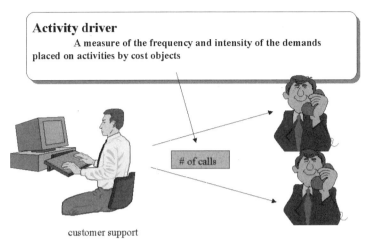

Exhibit 3.8 Definitions

HIDDEN BEAUTY OF THE CAM-I CROSS

When a business or enterprise is modeled using this methodology, users will gain significant insight into what work is essential and where money is being spent. Decision-making professionals can understand the following powerful concepts in the process:

- The relationships between the work performed (activities) and how it consumes resources
- The relationships between products and services created and activities performed
- What drives work that creates products and services (drivers)
- How to set performance targets and measures

Extensions to these basic ideas are:

- Contribution analysis: Working from specific products, what activities and resources contribute to it?
- Driver analysis: If these drivers are tweaked, what will happen to the volume of products?
- Activity analysis: If activities relate to each other, can their relationship be modeled?
- Profitability analysis: Which are the truly profitable products and services?
- Channel and customer profitability
- Should the firm make or buy?

- Is the firm charging enough for its services?
- Where should performance improvements be focused?
- How effective are promotional and marketing activities for new customers?
- Which activities support strategic objectives?
- What will happen if there was an increase in demand for the firm's services? Can it be maintained with limited resources?
- How can the current budget be justified? What are the impacts?

As the cross is traversed, several ways to view information from an activity-based lens can be seen. However, as the analysis is extended and the cross is rotated around any axis (see Exhibit 3.9), and as levels are changed from strategic to operational to financial questions, answers to multidimensional questions will be uncovered:

- What is the profitability by this channel, by this product?
- List all the contributors to cost that are non–value-added compared to last period.
- Does cycle time corelate to costs?

COMMON REALIZATIONS USING ABC AND THE CROSS

Many organizations discover that their highly complex products with low value or services are grossly undercosted and are subsidized by their high-volume, least complex products and/or services. (See Exhibit 3.10.)

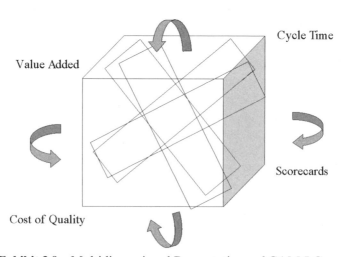

Exhibit 3.9 Multidimensional Permutations of CAM-I Cross

COMMON REALIZATIONS USING ABC AND THE CROSS 23

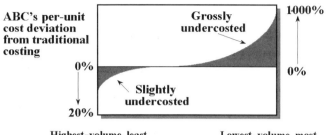

Exhibit 3.10   ABC Benchmarks

Many also find that approximately 70 percent of their products and services generate 400 percent of their profit while 30 percent take them back down to 100 percent. (See Exhibit 3.11.) Shockingly, diversity costs. Some products actually reduce profits. In some circles, product managers and business unit managers will justify this by stating that the products and/or services that do not make money are created for strategic reasons. This may be. However, companies have grown all too familiar with the notion that when things do not make money, they are called strategic!

Many decades ago Peter Drucker stated that 80 percent of a company's profits come from 20 percent of its customers. This realization is still true in business today. Also evident is the unfortunate predicament that organizations still cannot identify the main activities that contribute to the profit and to those that take profit away. ABC can provide this analysis.[7]

These two realizations are worth the exercise of ABC only if users can drill down under the skin to uncover the true illness—what is causing the symptoms. The next question is far more important: What do we do now? Here is where ABC shows its strength. With ABC, actions are part of the analysis

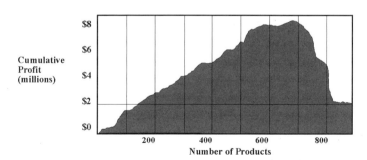

70% of the products generate 400% of the profit

Exhibit 3.11   ABC: Profit Justified

because users will be able to understand what resources and what activities cause these costs and losses.

Few analysis methods give operating professionals the depth of analysis that ABC provides. Again, organizations can pick up a strong grasp of the business activities and, in a way, rotate all analysis around these leverage points to uncover new insights. With a chart of accounts, financial teams have a handle on how assets are dispersed. In ABC, with a chart of activities, businesses can have a handle on how activities create wealth through products/services and how to harness them.

IS ABC JUST ANOTHER INITIATIVE?

Initiatives live fast and die easy. ABC almost fell into that chasm in the mid-1990s. Pitted against the general ledger, it may have died a harmless death but was reborn into the realm of decision support and analysis. As a methodology that is finance endorsed, operationally used, and information technology (IT) maintained, ABC is just coming into its own. ABC is hardly an initiative; it is a measurement underpinning to other initiatives.

Michael Hammer and Gary Hamel, the fathers of reengineering, defined business process reengineering to be the "fundamental rethinking and radical re-design of business processes to achieve dramatic improvements in critical, contemporary measures of performance, such as cost, quality, service, and speed."[8] Key words in this definition were:

- Fundamental
- Radical
- Processes
- Dramatic

The world embraced this approach. Today many believe that more than 50 percent of these initiatives have not lived up to their claim. As early as 1994, American companies spent approximately $32 billion on business reengineering; two-thirds failed. Yet it is useful to examine the reengineering promise, that dramatic results can be achieved by redesigning processes using contemporary performance measures. But many firms just redesigned processes to improve speed rather than looking at what to improve first, using all contemporary measures available. A key contemporary measure of cost is ABC. With ABC, users can focus their attention on areas of the business that need change. (Exhibits 3.12a,b and 3.13.) In many ways, ABC should be performed

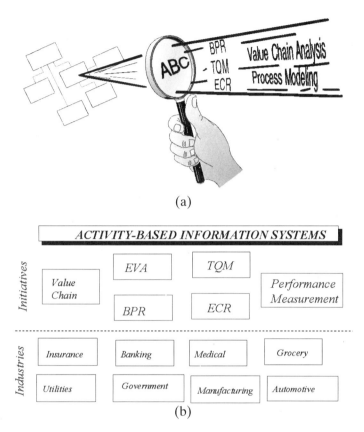

Exhibit 3.12 (a) Initiatives Need Strategic Focus and (b) Technology Enabler for Measuring Business Initiatives

before any other initiative so that organizations can learn where to target their initiatives.

COMPARISON OF FADS VS. EMBEDDED TECHNOLOGIES

Initially, ABC was viewed as a fad. Like other business ideas and techniques, ABC entered the business improvement market as a new and overwhelming idea. Pitted against the general ledger and other older cost management methods, it seemed revolutionary. In the early years, as a finance revolution, it took reasonable hold without enabling technology. Then ABC fell prey to pilot programs that could not continue. Today, with ongoing sustainable technology that enables continuity, ABC/M is moving from fad to a careful marriage between

26 EVOLUTION OF ACTIVITY-BASED COST SYSTEMS

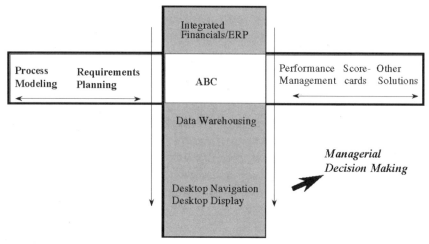

Exhibit 3.13 Beyond ABC

a finance concept and practical application of the domain using current technology. The following reminds us of the differences an organization has in perspectives, when engaging in an ABC/M endeavor:

Fads
- Limited life
- Are inoculations
- Use consultants and do not transfer knowledge
- One book away from latest craze in management circles
- Usually unintegrated information
- Part-time teams
- Leader is visionary
- Like quick-fix oriented
- Immediate value search
- Standalone
- Inexpensive
- Proprietary
- Closed-in architecture
- Limited service, support, and training
- Limited upgrades and updates to creation
- Frequent "ahas"

Embedded Technologies
- Are a way of doing business
- A regimen
- Competitive advantage

- Dedicated teams
- Integrated
- Leader is pragmatist
- A living "diet plan"
- Long-term value with immediate side effects
- Scalable and maintainable
- Higher price/performance
- Standards based
- Open architecture
- Strong service, support, and training
- Applications services component
- Regular upgrades and updates
- Frequent "I knew thats" as confirmation

Initiatives tend to be unidimensional and have the promise of a quick fix. ABC hardly fits this billing, although some organizations do design their ABC endeavors as a quick fix and a one-shot event. Naturally, they get what they set out to build. So, if designed as a quick-fix, ABC/M can be one and thus become an initiative.

The difference between ABC/M and other initiatives is found in two areas:

1. ABC is about managing the business using multiple dimensions: profitability, by products, by services, by resources, with drivers, and so on.
2. Technology exists to transform the viewing of the organization from a general ledger–centric orientation to a resource/activity/products/services/channels–centric view.

Technology tends to prolong knowledge. When systems exist to retain and transform data into information—to deliver it to the desktop—the probability of transforming what may begin as a fad to a way of doing business is greater. Truly integrated and continuous technology delivers significant reinforcement.

FOOD FOR THOUGHT: ABC IN THE FOOD AND GROCERY INDUSTRY

In November 1998, at 8:00 P.M. central time in every one of 2,390 stores, Garth Brooks performed a live broadcast. Wal-Mart calls it "retail-tainment."[9] The shopping experience is fast becoming an entertainment event just to attract more customers. An indication of how far the shopping experience has changed, this event characterizes an industry in transition and acute transformation.

The food and retail industries, once enabled by traditional trading partners—that is, manufacturers, retailers, wholesalers, and brokers—are having to rethink their value proposition to themselves and each other.

It used to be that every member of the value chain did their part and took their piece. Now everyone realizes that the end customer has options and overall control. What used to be "hunt-and-peck" selling to these consumers—that is, "Give them a sale and they will come"—is now transforming to the "lifetime value of a customer." From membership card discounts to Electronic Data Interchange (EDI) technology, the industry is affixed on winning the mind share of the customer.

"Realizing that Wal-Mart was taking their business away, the food industry trade associations got to take the fat out of the value chain, and one of the enablers targeted was ABC/M" states Karen Ribler, president of KJR Consulting, a Washington, D.C., consulting firm.[10]

Another example of a powerful application of ABC is titled efficient consumer response (ECR). This worldwide initiative was formed because of the increased competitive pressures on the status quo of manufacturers, retailers, brokers, and wholesalers. This $500 billion industry is targeted to improve their business model and their mutual partnerships to reduce $30 billion in costs in the next five years.

As a natural consequence, the stores here today may not be there in five years. The way people buy groceries may evolve, and even what and how the food industry sells will change. ECR, a global phenomenon, has recognized four best practices to embrace:

1. Category management
2. Electronic data interchange (EDI)
3. ABC
4. Continuous replenishment

Traditional partners in the business have to step up to contemporary ways of doing business to increase their downward spiral of profit margins. Companies like Kraft Foods, Giant Foods, SuperValu, and others have led this march to optimize their supply chain. ABC is now viewed as a conservative business practice; it is no longer considered risky and leading edge.[11]

The power of ABC has been proven, but attempts to standardize activities across the value chain have yet to take hold. However, several distribution and suppliers in the grocery industry have embraced ABC to understand their costs. Procter & Gamble is well known for its use of ABC in streamlining logistics. Coca-Cola Company, Nabisco, Kraft, H.E.Butt, SuperValu, Fleming, and Spartan also have been active ABC users.[12]

ABC/M provides a common language for measuring and understanding costs, resources, and activities. ABC/M has been extended to measuring the value-chain performance as well.[13] For example, Kraft Foods and SuperValu (which has initiated four such relationships) used ABC as a means of isolating costs across the value chain. Richard De Santa, author of an article on the subject, states, "The foundation of the venture, being implemented by corresponding cross-functional teams which meet regularly through video conferencing, is the mutual identification of cost factors across all supply and distribution activities, provided by Activity-based Costing models in place at both companies."[14]

The middle-line or expense-line management cannot make organizations gain market share. Yet the industry realizes that using ABC, profitability profiles for customers, vendors, and products are not one-dimensional. Armed with activity information, users can manage to profitability rather than blindly cut costs.

ABC HAS UTILITY

The utility industry is being transformed with privatization in the 1990s. Merrill Lynch published a detailed report titled *"Utility Industry—Armed to the Teeth"* highlighting the role of ABM in differentiating one utility from the other.[15] This document is a testament to the strategic, operational, and financial value of ABC to a changing market as it states boldly: "In our opinion, utility investors have never faced an environment where management skills, processes and practices will make as much difference as over the next 3–5 years."[16]

This analysis and conclusions indicate and reaffirm that ABC implementations and the use of ABC are key to the relative value of utilities. Merrill Lynch's conclusions of ABM demand attention:

- Across the industry, ABM is one of the most important improvement vehicles for management practice.
- The utility industry is in its infancy in obtaining the true value of ABM.
- Initial monetary cost and employee morale are factors.
- Potential advantages are extensive:
  —Improved budgeting
  —Increased accountability
  —Productivity gains
  —Greater understanding of the flow of costs
  —Line manager independence—tracking their own performance
  —Better handle on performance by senior management

The utility industry is undergoing a major transformation and deregulation. With more focus on competition and on customers, the industry is searching for ways to optimize and to gain profits. Utilities have realized that they have customers with diverse needs. Some are restructuring their business units to serve those needs along with unbundling their products and services to align these needs with business units. ABC has been seen as a way to measure the business unit performance.

It is not often that business models are completely changed; however, the utility industry has been faced with a drastic overhaul. Almost all elements of its business and managerial systems must catch up.

POWERFUL SIDE EFFECTS OF ABC

ABC is the workhorse of the future rather than a cost initiative of the past. As a cost initiative, comparing it to traditional GAAP-oriented standards would serve no purpose. Billed as an adjunct to the general ledger and GAAP, ABC/M is flourishing and assisting other initiatives that are starved for measurement grounding.

Activities are the language of business, and ABC/M provides a careful linkage from these activities to the resources and products/services a business creates.

ABC/M as an exercise has provided many organizations salient value in transforming culture while affecting the bottom line. Most ABC conferences balance discussions on model building and technology with the true leadership and team politics evident in any project.

Politics of Performance Management

The side effects of an ABC/M program include:

- Managers gain a clear understanding of their organization's contribution to revenue and expense.
- Multifunction teams developing ABC models and endeavors are focused teams bringing the knowledge of an organization to a common objective.
- The vocabulary learned in ABC is valuable as a means of communication. Line managers and operational managers learn this common language that bridges the ultimate chasm of information—bridging dollars to products/services and to the work performed.
- Many times line managers are seldom heard. They are frustrated about the standard information packets they receive from finance and claim that the numbers seldom reflect reality. An ABC/M exercise requires that teams gather data about activities and develop a business model(s) relat-

ing these activities with the resources and products/services. Here line managers and operational teams are given the opportunity to mold the information they will use. ABC/M systems actually reflect the state of the internal relationships and functions of the business. The information gained contains productive insights. Operational managers will use it to begin to dialog and challenge the information. These actions will result in valid, relevant information.
- All this talk about "What do you do?" and "How much time do you spend doing x?" drives efficiency-centric thinking in an organization. As long as something is measured, it will be affected.

ABC/M can play a clear role in performance measurement because it provides significant information on costs to any measurement system. Paul Bierbusse, senior manager at Ernst & Young LLP, declares that performance management "requires that you take corrective actions to close performance gaps." ABC/M systems combined with scorecards that capture and organize nonfinancial measures give momentum to the performance measurement industry.

ABC/M IS POISED FOR GROWTH AND ADOPTION

Markets grow and take off not just because of great ideas applied to great need. Markets grow because all the preconditions for growth converge. The activity-based marketplace has enjoyed growth since the 1980s but the conditions of strong growth may be aligning themselves. Signals of change are clearly on the horizon. They are:

- The thought leaders are endorsing ABC as a critical part of the business landscape.
- The technology has achieved critical mass, breadth, and depth.
- Best practices are emerging and a body of successes and errors are documented.
- The community of practitioners is increasing.

Thought Leadership

Thought leadership and documented knowledge have spread since the 1980s. Independently, many leaders advocated changes that have transformed the landscape. Dr. Steven Covey advocates that ABC/M will "empower organizations with solid information about their organizations that enables them to exercise leadership and wisdom in decision making," while Peter Drucker, the father of modern management, has gone even farther, declaring that those who do not accept and acknowledge ABC in the year 2000+ beyond would be obsolete.[17]

Technology Momentum Drivers

The market has absorbed the introduction of several professional ABC/M software organizations that have brought breadth and depth to the technology. Other technologies, such as on-line analytical processing (OLAP), multidimensionality, profitability, activity-based budgeting, on-line surveys, and rules-based model checking are but few examples of depth. The market has expanded horizontally to incorporate process mapping, balanced scorecard, and several connections to the enterprise resource planning (ERP) systems. (See Exhibit 3.13.)

ABC/M technology practitioners have grown with experience. As the technology becomes more accessible and understandable, mainstream business managers will begin to find this technology on their desktop and begin to demand more conservative and pragmatic demands of repeatability, scalability, and enterprise-wide deployment. The computing industry is finally at a stage of development, integration, and standards that information technology can deliver. Via high-powered data warehouses, business modeling tools, and desktop navigation tools, the promise of ABC information on the desktop will be satisfied.

These technologies are now deeper and broader than ever before. With the advent of the new ERP firms, all introducing ABC systems, the market has been legitimized. Combined with data warehousing, data mining and agent technologies, emphasis is being placed on the technology driven activity-based information on the desktop. The availability of desktops and the clustered power of workstations have fueled further growth and desire for information anywhere-anytime.

Best Practices Are Emerging

CAM-I has produced several focused studies in ABM starting with early development of the glossary of terms.[18] The American Productivity and Quality Center (APQC) with CAM-I and Arthur Andersen & Co. LLP joined forces to investigate and develop an understanding of best practices.[19] Countless books and videos spanning several industries, from government, to process industries, to manufacturing industries, have been created.[20]

Community Is Learning the Market and the Skills

Conferences, books, videos, technology seminars, and consortium gatherings have been gaining momentum. Articles and books litter the landscape, and the ABM community is growing at fast pace. Practitioners who have gained the knowledge of experience tend not to stay in the same company. Some tend to move to another to implement ABC again, thereby increasing and training new recruits in the field.

WHERE DO WE GO FROM HERE?

From 1980s to the early 1990s, ABC/M was viewed from the lens of finance. With this perspective, it served a purpose of cost management and effectiveness—namely accounting. The inherent cross-functional value of ABC is now being harnessed; the genius of its simplicity is being applied to the overall business, beyond overhead to the way businesses are viewed. Today, a fair profile of a generic business adopting ABC/M for their business is this:

Companies that employ ABC have something to fix—expense control related, operational product costing, whatever. They are in need. Many companies that are enjoying record profits and are drunk with success seldom consider ABC. Finance organizations usually generate ABC/M. Usually delegated by the CFO, ABC becomes the sole project of a project leader reporting to the CFO.

Approximately 50 percent of firms use consultants. Others go at it with in-house knowledge and talent. The market was 100 percent consultant driven in the early years, but this has changed as more and more successful practitioners are being generated through knowledge transfer and proliferation.

Multifunction pilot teams form the first ABC projects in Fortune 100 companies. Steering committees also sit above these teams to guide them. Many organizations begin with a proof-of-concept exercise. They are usually chartered to solve specific product/service problems. Their mind-set is visionary, and they seek to gain what is called the "aha effect"—they are looking for dramatic realization rather than to confirm their gut feel.

Many pilots are one-shot events, dying after they succeed. Others expand to be multisite programs. Fewer still become truly enterprise-wide deployments. More details on the nature and personality of this growth are provided in Chapter 14.

Now that we know what the general personality of ABC/M project is, the momentum drivers we discussed earlier are transforming and accelerating the nature of these projects. In a nutshell, we are moving from the above-mentioned personality to include a deeper more lasting scenario:

Moving from	To include
Pilot/proof of concept	Enterprise-wide activity management systems.
Finance born/initiated	Finance endorsed, operationally used, IT maintained.
Initiative or fad	Workhorse, always around to add to the business.
Consultant driven/performed	Consultant guided, internally driven.
Part-time teams	Dedicated full-time, multifunction work groups
Accounting management	Accountability management
Looking for the "aha effect"	Looking to validate the belief systems of operations, the "I knew it" effect

UNDERSTANDING THE POTENTIAL OF ABC/M

An understanding of what ABC can do for an organization is the first step to winning with ABC. Traditionally, ABC is about costs, but as an ABC/M project matures new applications of the information surface. ABC champions can now move to understanding the phases that new entrants follow in learning and building an ABC/M system.

NOTES

1. Some treatment of the topic can also be found in J. Miller, and T. Vollman, "The Hidden Factory," *Harvard Business Review* (September-October 1985).
2. Norm Raffish, and Peter B. B. Turney (eds.), *The CAM-I Glossary of Activity-Based Management, Version 1.2* (Arlington, TX: The Consortium for Advanced Manufacturing-International, 1992). The glossary is reprinted in the back of this book.
3. H. Thomas Johnson and Robert S. Kaplan, *Relevance Lost: The Rise and Fall of Management Accounting* (Boston: Harvard Business School Press, 1987).
4. J. Miller, *Implementing Activity-Based Management in Daily Operations* (New York: John Wiley & Sons, 1996); M. R. Ostrenga, Terrence R. Osan, Robert D. Mcilhartan, Marcus D. Harwood, *Ernst & Young Guide to Total Cost Management* (New York: John Wiley & Sons, 1992).
5. Note that there are two distinct ways to model in ABC: One is the CAM-I method and the other is the bill-of-costs method (sometimes called the

output measure methodology). The latter can be a far more effective and less complex method to model and simulate an enterprise. Both methods have their advantages and disadvantages. The output measure method removes a significant amount of complexity in the traversal algorithm. For the sake of simplicity, the author has chosen the CAM-I methodology as it is the more understood in the United States and seems easier to grasp for the beginner. Please note that the output measure methodology is worthy of attention and respect.
6. Raffish and Turney, *The CAM-I Glossary.*
7. Tad Leahy, "Making Sure the Customer Is Always Right," *Business Finance* (October 1998), pp. 61–63.
8. M. Hammer & J. Champy, *Reengineering the Corporation* (New York: HarperBusiness, 1993).
9. Emily Nelson, "Wal-Mart's Garth-Quake May Spur Sales," *Wall Street Journal,* November 2, 1998, p. B1.
10. Tad Leahy, "ABM—When to Hold 'em, When to Fold 'em," *Business Finance* (November 1998), pp. 53–54.
11. Carol Casper, "A Value for the Value Chain," *Food Logistics* (June/July 1997), pp. 30–32.
12. Michael Garry, "ABC in Action," *Progressive Grocer* (February 1996), pp. 71–72.
13. Casper, "A Value for the Value Chain."
14. Richard De Santa, "Real-life ECR—A Most Measured Approach," *Supermarket Business* (November 1996), p. 16.
15. Stanford Cohen, Doris Kelley, Daniel Ford, and Mary Galvin, *Utility Industry—Armed to the Teeth* (New York: Merrill Lynch & Co., 1994).
16. Id., p. 6.
17. S. Player and D. Keys, *ABM: Lessons from the ABM Battlefield* (New York: Mastermedia Ltd., 1995). 1997 Cognos Corp. Users Group meeting where Drucker gave the keynote address.
18. See *www.cami.org* and CAM-I Glossary.
19. Steve Player, "The ABM Tidal Wave," *Controller Magazine* (December 1997), pp. 71–72; Contact Arthur Andersen & Co. Cost Management Group for more information.
20. See *www.abctech.com* under ABC University for literature and books. Use www.wiley.com as well for searches.

4

ACTIVITY-BASED MANAGEMENT LEARNING LIFE CYCLE

Many ABC/M endeavors go through certain growth and learning phases. Almost 50 percent of all endeavors, however, fail to reach their true potential; in other words, they never go beyond piloting.

There are many reasons for this, but one of the main reasons is that companies take on too much too fast and do not respect the elements that are needed to build an endeavor. Parker Hannifin, a Cleveland-based manufacturing company, views ABC as "a way of life." Fred Garbinski, assistant controller of Parker Hannifin, advises people not to try to implement ABC systems organization-wide from the start. He feels it does not work as people do not understand it quickly. Rather he recommends a "slow, methodical, building block approach."[1]

Four distinct phases of the life cycle of an ABC/M project have emerged through the years. (See Exhibit 4.1.) These phases are:

1. The trigger phase
2. The education phase
3. The pilot phase
4. The enterprise phase:
   - Local
   - Global
   - Virtual

TRIGGER PHASE

Most ABC programs are launched to solve a set of problems. Very seldom is ABC/M employed as a natural way of doing business—just to improve it. "Business units fail to re-focus because they are pre-occupied with the present

TRIGGERS		Pilot	Local	Global	Virtual
	EDUCATION				
Software	No software Needs analysis	PC-based Software	LAN-based software with local deployment	Client-server software with global piloting	ERP systems Analytic systems
Resources	Consulting, learning, problem bounding	Classroom Training On-site Training Technical Support Rapid Prototyping User Group Books Videos Conferences	Classroom Training On-site Training Technical Support Application Services User Group Books Videos etc.	Classroom Training Web-based/ On-site Training Technical Support Application Services User Group Books Videos etc.	Classroom Training Web-based/ On-site Training Technical Support Application Services User Group Books Videos etc.
		ANALYTIC		ANALYTIC/OPERATIONAL	
				ENTERPRISE	

Exhibit 4.1 Phases of ABC/M Implementation

or the past. Success is a double-edged sword," states John Whitney, professor of management at Columbia University's School of Business, when discussing how success in the past or present can "lull a company into complacency."[2] As ABC/M is more understood, more organizations will adopt its methods as a natural course of business. Today it takes a "pain" of some sort for ABC/M to be adopted. Some event—competitive pressures, a need for better and more accurate pricing, a loss in momentum, or an actual loss in revenues or profit—can form a trigger.

In these cases, the shock usually widens people's pupils and they tend to search for ways to improve. Characteristically, most companies in this situation look for the quick fix and resort to ABC/M for support.

The following list contains examples of triggers.

Industry	*Market and Internal Triggers*
Medical	Managed care/cost of delivering services
Utilities	Global privatization
Government	Doing more done with a smaller budget
	Budget justification and impact statements
	Outsourcing
Banking/Finance	Consolidation and acquisition
	Optimization of the cost of delivery
Groceries/Food	Efficient consumer response

All in all, external forces create the need for considering new methods of cost management. In the end, certain conditions foster the invitation of ABC/M as the primary activity.

One cautionary word: Often at this stage ABC/M is viewed as the answer to all ills. Teams and champions tend to promise more than can be delivered. By so doing, they set in place the conditions for failure.

EDUCATION PHASE

Born in finance and accounting, ABC/M endeavors are becoming more and more mainstream only in the minds of ABC/M aficionados. A much larger group of people exist within an organization who must be educated in ABC/M. ABC/M champions sometimes forget that they must educate the vast array of people back in the office. Frustrated with the lack of acceptance from their organizations, the champions of ABC often think them behind the times.

This educational phase forms the foundation necessary for the organization to accept and employ ABC. In this introductory phase, champions and teams

PILOT PHASE

are asking only one question: "What it is?" Other questions, such as "Will it work?" or "Will it work continuously?" take second and third places.

Those who watch several ABC implementations could conclude that the implementation curve from pilot to propagation is directly proportional to the learning curve within the organization in question. If an organization teaches the value of ABC/M and its people really learn, the adoption curve of ABC will be accelerated.

Often underestimated, the education phase of a project is usually done quickly and restricted to the project team. However, formal educational programs are a necessary and sufficient condition for project ignition and success. Saturn Corporations' CEO, Richard G. LeFauvre, states "If you think education is expensive, just try ignorance."[3]

The education phase is one that never ends in an ABC/M program. The audience of learners will grow if the project scope grows. Ashok Vadgama, executive in charge of ABM at Motorola, started his project with education in mind. He put programs in place to grow his audience of users. Now he maintains a Web site accessible to many decision makers that updates them on the mainstream project for which he is responsible. With the click of a mouse button, interested parties can learn about ABC/M as well as read reports and results derived from the ABC/M exercise.[4]

All in all, this phase is the most important and should be planned carefully because every other phase depends on this being executed well.

PILOT PHASE

Once in pilot programs, ABC champions and teams are trying to prove the concept of ABC and also testing the organization's ability to accept the methodology. In the middle of the evolution curve, multifunction teams usually form. These teams focus on:

- Self-training and education
- Development of basic models
- Using stand-alone modeling environments, usually PC-based
- Being guided by external consultants or educators
- Searching for the "ahas" to impress management with the value of the team
- Selling management and the operating teams on the value of the program

These teams are trying to answer the basic question of "Will it work?" more than any other. Note that some pilots almost look like enterprise rollouts because they have global deployment implementations; they have multimodeling teams across many regions and seem to use the information at the operational level. They are still pilots because they are not the mainstream business method. Pilots have certain characteristics. They:

- Tend to be short and make a point
- Are encouraged by management rather than expected by management
- Behave like other initiatives with a lot of dust in the sky and loud fanfare
- Live fast and die with no one person fully dedicated to them
- Have a 50-50 chance of survival
- Have a half-life of two years

On a cautionary note, members of a pilot team should be sure to test not just the system concept but the:

- Acceptance of the concept
- Ability of the organization to understand and engage on the topic
- Ability of the organization to learn
- Ability of the organization to accept the different information
- Technology absorption capability of the operational teams
- Technology absorption capability of the management teams
- Interface demands awaiting if the project would go live

Most important, it must be ensured that if all goes well, someone makes decisions with the information and does something. The most frustrating challenge to an ABC champion is to accept that a great project ended in no decisions and changes. Anticipating this, the best way to teach an organization to respond is to provide mock-up reports of information and test reactions by asking "If I got you this, what decisions would you make?"

For example, one consultant delivered a stack of ABC/M reports to her clients to introduce them to the concept of ABC. Then she would ask them what they would like in the ABC reports. With this information, she would build their models. A further step would be to ask what decisions they would make if confronted with these reports.

The writers and consultants James Collins and Jerry Porras state that "having a great idea or being a charismatic visionary leader is 'time telling'; building a company that can prosper far beyond the presence of any single leader and through multiple product life cycles is 'clock building'."[5]

In the same way, pilot programs are time-telling exercises—they prove the concept and test the possibilities. Production system implementation is the proof of clock building and the test of the realities created through the dreams and visions of pilot programs.

ENTERPRISE PHASE

The enterprise phase is a natural consequence to the successful pilot. However, as many as 50 percent of pilots never go to another phase; they either die or remain in limbo. One reason for this is that many companies do not anticipate the technical requirements of this phase when doing a pilot.

In a pilot, systems are built for the "quick hit." The average time between initiation and results is short, and the requirements for accuracy, repeatability and maintainability are low. Many times ABC/M project leaders cut corners to achieve the "big bang" they are seeking. After the applause dies down and the "plaques" for victory are distributed and hung, someone on the team is usually asked to make this work all the time with the same temporary resources. The second time is less exciting, and the champion is unable to hold his breath as he did before.

The demands of enterprise deployment include:

- Regular data integration demands
- Regular data-gathering methods for both nonfinancial and financial information
- Regular ways of getting empirical information (information found in people's heads)
- Custom reports developed and adjusted to the needs of operational teams and managers
- Constant education to all concerned on ABC/M
- Regular orientation to new participants
- Competency centers to assist any user of computing systems
- IT involvement and recognition of the need for standards to protect the investment
- Interest in automation because manual translation of information is challenging

For ABC/M, production systems can be varied. When champions talk of their implementations, the same words sometimes mean different things. Enterprises could mean:

- We have built a pilot model of a P&L and are now going to expand that model to a larger enterprise.
- We have completed a basic model for a site and now we shall build other site models.
- We are now ready to present our model to the operational teams to move from ABC to ABM, where decisions are made with analysis of results.

- We want to model our entire enterprise across a multinational company.
- We want to deploy activity information across the enterprise moving data that are financially endorsed, IT-maintained, and operationally used by decision makers using desktop tools.

Clearly, some of these enterprise deployments are local (within a certain country or region) while others are global (among countries and nations) and they need the involvement of different teams in implementation. Exhibit 4.2 shows a globally deployed ABC system. Such a system requires significant enterprise deployment control systems. The true global deployment demands significant resource commitment and checks-and-balances or else the same mistake will be repeated across the globe; some design controls are needed on models created across the globe. Each site, if left to itself, will create its own model in its own way. When a consolidated view of the models is wanted, trouble begins. However, if a team controls everything in the structure and the data of the models, the various cultural and location specific anomalies in a company's architecture will never surface and the models will not reflect reality.

Some suggestions are:

- Force a standard dictionary for models: for example, outline what activities are available to be identified and what their characteristics are.
- To allow for cultural differences across regions, allow for certain variations in the model design and ensure that sign-off or certification occurs.

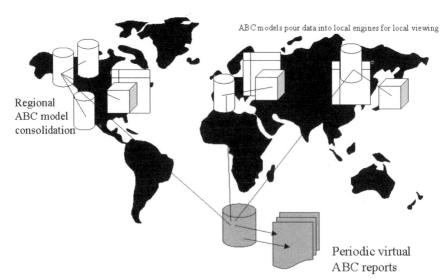

Exhibit 4.2 Enterprise Deployment

ENTERPRISE PHASE

- Develop manuals for model creation and reporting specific to the business.
- Train a competency center to support the systems being created and to be a clearinghouse for consistency.
- Have a champion travel to each site to ensure problems are removed or solved. Sometimes this champion travels to each plant and ensures that models are developed consistently. While this is one way to ensure consistency, be aware that only this person knows how and what has been created.
- Test all technologies to consolidate models, link this ABC/M information, and ensure consistency across the enterprise before going into implementation.
- Ensure that all sites report regularly—once a quarter—and ensure that the model can be refreshed once a quarter at least using the most up-to-date processes.

Enterprise systems graduate from local deployment of information to operational teams, as proof-of-concept, to global deployments.

When ABC/M stand-alone or client-server systems are used to enable enterprise deployment, they are still proof-of-concept experiments.

Once these systems are linked to existing operational and/or integrated ERP systems and are regularly performing parallel functions, virtual enterprise analysis and deployment are achieved.[6]

Stand-alone ABC/M systems perform two functions in these phases:

1. In the beginning, these systems are excellent prototyping, modeling, and analysis proof-of-concept systems.
2. After the ERP systems and these stand-alone systems are connected, the ABC/M systems perform essential analytic and modeling functions almost like a connected yet off-line engine that can continually test and develop new ideas while the production on-line ERP systems serve the enterprise.

Even though ERP systems have ABC modules integrated in their offerings, this does not eliminate the essential analytic applications that remain connected but external to the ERP system. Hence, stand-alone ABC/M systems are called analytic applications.

Enterprise-system implementations of ABC/M are still not common. In fact, claims of stand-alone PC enterprise deployment of ABC/M information are premature. Until ERP and analytic stand-alone systems work in concert with each other across the global enterprise, true operational deployment of information will not be a reality. ERP and stand-alone ABC/M systems can provide both the analytic and the integrated operational needs of an enterprise. With this, activity-based information systems encompass these two, working with data warehouses and desktop navigation tools to gather and provide ABC/M information to teams.

Hence, claims of enterprise ABC/M using stand-alone PC technology are actually attempts at educating, training, modeling, and deployment in the prototyping phase of a project. At this pre-ERP phase, organizations have the opportunity to test their hypotheses and methodologies with live users. When both ERPs and analytic ABC/M tools work in concert to model and deploy, respectively, activity information, ABIS will be achieved.

New technologies are being introduced daily, thanks to the companies, such as Microsoft Corporation, that are making the networks of tomorrow almost transparent. With the advent of intranets and extranets, information will be delivered to users and accessible without any knowledge of where the information is physically located. Users will be talking to a server in Japan and may not need to know about that.

Consequently, the future ABC/M environment will be almost virtual; knowledge on the desktop will appear via a simple browser accessing model information from several consolidated models across the universe of models in the company. Thus far, ABC/M environments are based on pull technologies—information has to be pulled out to the desktop by managers. These managers then review the information and interrogate the data. In the future, information technology will be pushed, that is, the manager will be able to set personality requests on the system and allow the system to collect information under the specified conditions. Disregarding all other information, certain ABC/M conditions will be recorded and the manager will be informed. The systems will understand users and their personal needs.

This virtual enterprise phase is around the corner and within grasp of the leading technology companies in ABC/M.

NOTES

1. I. McLemore, "Overhauling Cost Accounting Systems," *Controller Magazine* (November 1996), p. 46.
2. J. Whitney, "Strategic Renewal for Business Units," *Harvard Business Review* (July-August 1996). pp. 84–98.
3. Bala Balachandran, "Cost Management at Saturn: A Case Study," *Business Week Executive Briefing Services,* Vol. 5, 1994, pp. 25–28.

4. Based on Ashok Vagama's presentation at CAM-I Fall 1998 conference.
5. J. Collins and Jerry Porras, *Built to Last* (New York: HarperCollins, 1994), p. 23.
6. Technology Marketing, SAP AG "R/3 System; The R/3-OROS Integration," SAP, October 1998, p. 5.

5

WHAT IS AN ACTIVITY-BASED INFORMATION SYSTEM?

Three classes of activity-based information systems exist:

1. One integrated into legacy, accounting, and resource planning systems
2. A best-of-breed analytic application used alongside integrated accounting systems
3. A hybrid combining both integrated systems and best-of-breed technology

The overall architectures and trade-offs associated with these systems are discussed in Chapter 6. Here we discuss the various subsystems that make up an activity-based information system (ABIS).

In contrast to an ABIS, an activity-based costing (ABC) system merely models the flow of costs from resources, activities, and cost objects. An ABIS is a set of software programs running on general-purpose hardware (PCs or otherwise). This system models the effect that activities have on a business and understands cause and effect using activity-based costing as its backbone. Using cost modeling, an ABIS can measure and model the relationships among resources, activities, and cost objects (i.e., customers, channels, products, and services). Furthermore, an ABIS assigns performance measures to activities, products/services/channels, attributes or tags them with both nonfinancial and financial personalities such as value and cost of quality, thus creating a truly multidimensional map of the enterprise. An ABIS is a conceptual framework embodying both an ERP and an analytic system.

This database of relationships is made of the following conceptual subsystems (see Exhibit 5.1):

- Data collection and input subsystem
- Modeling and analysis subsystem

DATA COLLECTION AND INPUT SUBSYSTEM

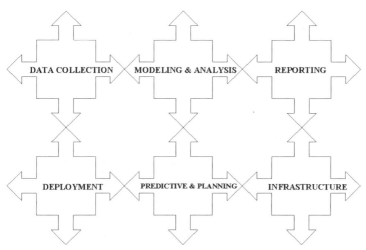

Exhibit 5.1 What Is an ABIS?

- Reporting subsystem
- Deployment subsystem
- Predictive and planning subsystem
- Infrastructure subsystem

DATA COLLECTION AND INPUT SUBSYSTEM

The importance of data collection is understood only after the first activity-based costing/management (ABC/M) exercise. At the pilot, the information gathered through interviews, while tapping ASCII data dumps from heterogeneous databases, seems trivial because the excitement is high. Everyone is willing to spend time and money on a first try. Everyone is also willing to hard-code the connections with data and not wonder what to do when the data structures change.

When model data and structures must be replenished continuously, the real challenge to the maintainability of the system surfaces. Many ABISs provide data collection tools, but few provide electronic data replenishment (EDR) tools (tools designed to assist in repeated data collection of resource data, activity data, driver information that is found in people's heads).

EDR is not focused on collecting data for the first time. Many other methods and systems allow for this. The biggest challenge in data collection is in data replenishment—gathering data the second and third time from the same peo-

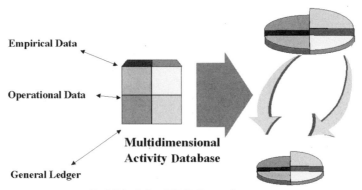

Exhibit 5.2 EDR Data Sources

ple. Data are systems-collected information but also information found in people's heads, that is, activity information or percentage of time spent information, often called empirical information, is essential to the process. Model builders spend a considerable time collecting data and interviewing people to get this information only found in the heads of the people doing the work. Technology is needed to facilitate gathering of both types of information. Exhibit 5.2 illustrates this variety of information pushed into a data warehouse or a data navigations tool.

Input and output subsystems take the following forms (see Exhibit 5.3):

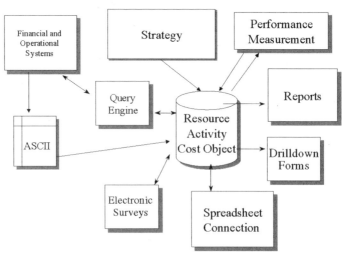

Exhibit 5.3 Data Input Stream and Methods

- ASCII importing and exporting
- Object data base connectivity (ODBC) connections with a query engine (EDR tools)
- Surveylike Web-based input and output systems (EDR tools)
- Excel-like front-end input systems

Some installations have as many as 80 connections to heterogeneous data sources. In certain organizations the data is found only in people's heads. Service companies, for example, tend not to have activity driver information logged in systems. Manufacturing companies that have production systems tend to have some manufacturing related driver information on-line. The data collection can be vast or very targeted. It all depends on the expanse of the data model being created. Setting up a system for a project with 1,000 cost objects and 400 drivers with 2,000 activities will take longer than for a project consisting of 10 drivers and 10 activities.

Exhibit 5.3 highlights graphically the data sources and tools while Exhibit 5.4 describes their trade-offs and value to the data collection process.

MODELING AND ANALYSIS SUBSYSTEM

As CAM-I standards would expect, any ABC modeling and analysis subsystem must be capable of capturing the CAM-I cross and its data elements. (See Exhibit 5.5.) A relatively unknown fact is that there are two ways to model in ABC:

1. The CAM-I way, which pushes costs through a cost assignment view—from resources, to activities, to cost objects. This method is the most popular in the United States.
2. The output measure methodology is more of a pull, where costs are pulled from a bill of costs. More intuitive to manufacturing companies familiar with bill of materials (but applicable to any industry), the output measure methodology does not use assignment paths like CAM-I but reduces network paths with a set of bills associated with each resource and cost object. This method is a very important and significant approach for activity-based budgeting and predictive methods used in more advanced installations.

Modeling and analysis subsystems must be able to offer both views. In order to thoroughly capture and evaluate the enterprise or pilot environment, a peri-

Interfaces	Purpose	Advantages	Disadvantages
ASCII	Raw input of importing, exporting. ASCII data files in specified format.	Quick access. Simple interface and defined among many software packages	One time only It. must take all the ASCII data.
ODBC QUERY	Standards-based query engine that can grab information selectively.	Very directed and selective connection to data. Very scripted and repeatable.	Must learn the query engine language.
SURVEY TOOLS	Empirical knowledge—information not found on systems can be captured.	Gets to information found in people's heads, not systems. Is anonymous and increases data correctness. Repeatable and measurable as you can find out how many really responded and urge more response. Less time consuming than interviews.	Must understand another feature of the system to collect information. Surveys can be intimidating to noncomputer-literate users and may not get all the salient effects derived from eye-to-eye interviews.
SPREADSHEET	Simple input mechanics.	Nonintimidating input and output. Graphical connections are known and understood.	Too simple. Cannot collect vast amount of data as spreadsheet runs out of steam with limited storage capabilities. Data collection is unidimensional unless time is spent to create multiple dimensions.

Exhibit 5.4 Trade-offs of Each Data Source Gathering

MODELING AND ANALYSIS SUBSYSTEM

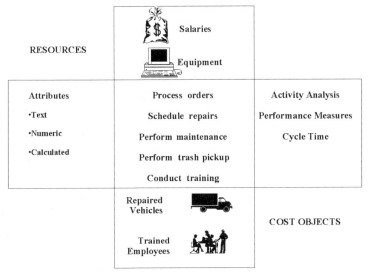

Exhibit 5.5 ABC Methodology

odic snapshot of the flow of costs through the resources, activities, and cost objects is essential and forms the basic building block for the development of a model.

As discussed in prior chapters, many organizations are satisfied with cost modeling at first but, very quickly, realize that their model needs to be extended to measure not only costs but revenue and profits as well. Some consider the model to be a measurement backbone for other initiatives or other measurement subsystems. For example, total quality management (TQM) projects often wish to know the cost of quality in their exercises. Process improvement systems also need accurate ABC rates to determine, for example, how much activity A costs.

Hence ABIS can take and have taken a deeper role than basic modeling of costs. Nevertheless, this basic view of ABC can have major implications to overall measurement initiatives.

In a nutshell, the modeling and analysis subsystem can create several views of the data:

- The classic ABC/M model and its cost-oriented structure
- The process views showing the relationship of activities to each other and their parents, namely tasks and processes
- The multidimensional views such as profitability by channel or by product or many more—a slice-and-dice model

- The scorecard view that uses this base to reflect performance metrics held by the organization

REPORTING AND DEPLOYMENT SUBSYSTEMS

ABC projects seem to always conclude anticlimactically. At the end of an exercise, people are presented with a rate: How much does it cost to do x? This fundamental critical exercise of discovering the true cost of services, product, channels, and activities can be viewed as the finale for the first-stage project. However, by using this information as a starting point, teams gain an enormous understanding of their own company from the inside out and begin to ask the questions that place them on the road to improvement.

Key to this learning is the "report-out" meeting held to inform management of findings. All multifunction project teams conclude their work with presentations that report on the model's output and respond to queries about the work performed. These meetings seem always to occur too early.

Reporting subsystems are critical at this moment. They produce two-dimensional reports with graphs that bring the entire data gathering, modeling, and analysis exercise into perspective.

Today there are two ways to report on the project data: (1) built-in reports from the software and (2) custom-built computer-derived files that permit personal, multidimensional computer investigation by management.

The built-in reports are custom reports designed for the special organization expectation. Some custom report writers also have a built-in query engine that grabs information just as a data collection engine does. These engines allow for custom reports to be built by designing reports with links to data elements within the ABC/M model. Then users merely run the engine to be updated on the status of the information. Further, enhancement to this approach touches the fields of data mining—often called "report mining." Here scripts can be set up to awaken the query engine, interrogate the models, and invoke the reporting subsystem (see Exhibit 5.6) when certain data triggers occur—for example, if costs exceed $5 million for product x.

The second type of reports uses data navigation software to drill down and analyze the data. These tools are also called EIS (executive information systems). These tools, a refreshing adjunct to flat reporting, permit users to interrogate the data and navigate through a succession of graphs, drilling down to uncover the true causes of cost or other dimensions in the model.

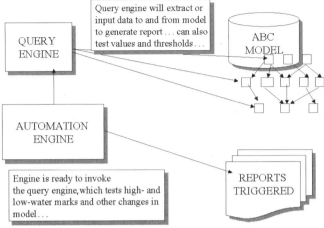

Exhibit 5.6 Report Mining

PREDICTIVE AND PLANNING SUBSYSTEMS

C. J. McNair, professor of management accounting at Babson College, states "knowing the score isn't the goal—changing it is."[1] In the end, it's not the information on the desktop that is key; it is the accuracy, timeliness, and applicability of the decisions made from it. Managers deal with the present using data from the past. Inherently, they are asked to predict the future, anticipate responses, and create them. Thus far, the tools available have been limited and not everyone has the time, the training, or the energy to develop elaborate models of reality.

However, the ABC/M framework provides some strong insight into performing rudimentary activity-based predictions and planning. Some opportunities are:

- Activity-based budgeting
- Target costing
- Balanced scorecard and performance measurement
- Process measurement and tracking—time-based simulations
- Capacity planning
- Yield analysis and predictions
- What-if analysis

All of these requirements must be reflected in any ABI system under the predictive subsystem. Although not all of these features are found in every software information system, most are migrating in these directions.

INFRASTRUCTURE SUBSYSTEM

In more advanced ABI systems, infrastructure holds the disparate tools in place and permits users to manage the information flow through them. Infrastructure controls the licenses in the toolset and restricts access at various levels of the software. Some packages are stand-alone PC tools while others are LAN-based (local area network) network accessible frameworks that hold and monitor access to tools.

Issues of security, licensing, and network capabilities are key concerns to most corporate system administrators and IT professionals. No two ABISs are built alike with the same administrative paradigm in mind. They must always be built on general-purpose computing environments; an IT-professional can assist in determining exactly what is needed.

One way to get a start on this is to ask the following questions:

How will the information system be used—from a stand-alone computer or on a network?
How many people must have access to the system and at which phases of the project?
What are the key tools needed to get the job done?
Who will maintain the system and help fix computer-related problems within the company?
Does this tool need to be a client-server? LAN-based network? Or stand-alone?

VIRTUAL NATURE OF SUBSYSTEMS

These subsystems can exist on multiple systems with heterogeneous base architectures and databases. For example, the organization may have an enterprise resource planning that handles product costing and operational reporting while certain elements of predictive ABC modeling and profitability modeling are handled by an off-line, connected PC-based system. Note that both these systems are valuable. Clearly, the stand-alone analytic systems can train, educate, and premodel the ABC/M environment; the ERP systems can collect data, deploy operational reports, and manage the major elements of the organization. Both of these systems comprise the ABIS in this context. These systems combine the characteristic efficiencies of their software to provide value. When combined with sophisticated data warehousing technology and desktop navigation tools, these systems can enable organizations to achieve true business intelligence.

NOTES

1. C. J. McNair, "To Serve the Customer Within," *Journal of Cost Management* (Winter 1996), p. 42

6

ULTIMATE PARTNERSHIP: ACTIVITY-BASED COSTING AND BUSINESS INTELLIGENCE

According to Richard Connelly, Robin McNeil, and Roland Mosimann, authors of *The Multidimensional Manager,* a new breed of managers is emerging—the multidimensional manager.[1] This is a manager who has moved from an intuitive style to a balance between intuition and analytic-centered framework. This manager has replaced the traditionally emphasized product and revenue mix to a more customer- and profit-centric model of thinking and working. Connelly, McNeil, and Mosimann claim that "more than 250,000 managers and professionals have become multidimensional managers. Their numbers are doubling every year."[2] The software and systems that enable such a manager are "business intelligence" software: software that brings knowledge to the desktop using tools that allow for analysis and navigation of multidimensional information.

Howard Dresner, noted analyst at the Gartner Group, adds: "Instead of small numbers of analysts spending 100% of their time analyzing data, all managers and professionals will spend 10% of their time using BI (Business Intelligence) software."[3]

THREE APPROACHES TO ABC/M SYSTEMS

In the context of business intelligence software systems, ABIS provides three distinct views to organizations. Thought leader and author Steven Hronec identified three views of cost management:

1. Financial: historical financial statements compiled to external rules
2. Operational: cost of running the business on a day-to-day basis
3. Strategic: analysis used to support long-term decisions[4]

Robert Kaplan and Robin Cooper state that one single system to serve the strategic and the operational would be inadequate.[5] They viewed the operational as "performing activities more efficiently," that is, doing things right, and the strategic as "choosing the activities we should perform," that is, doing the right things. Hence, under their definition, business actions that deal with quality, operational outputs, performance measurements, and reengineering would fall into operational frameworks. Business actions that are predictive and planning oriented, that is, segmenting markets, customer/product/service/channel modeling and analysis, would fall into the strategic framework.

Steve Player, firmwide partner for Cost Management at Arthur Andersen & Co. LLP, adds "The simple truth is that it's very difficult for a single system, even an activity-based one, simultaneously, to meet the requirements of all three of these different views."[6] Kaplan and Cooper elaborate that "Activity-based cost systems give managers a more strategic view of their business by helping them understand the sustainable economics of making products and serving customers."[7]

Although hard lines can be drawn between physical systems, isolating one as analytic and one as operational, subsystems can exist that share information and even functionality. Furthermore, although Cooper and Kaplan believe that operational and strategic systems cannot be one and the same, there are others who believe that they can.

For this discussion, consider that subsystems may share with one another across physical boundaries to make the entire collection of subsystems form an activity-based information system. Also note that analytic systems serve a fundamental need, while integrated operational systems serve another. Analytic systems currently center around PC-based stand-alone systems, which are easier to model and get up to speed than ERP systems, which cover a wider range of the solution for companies.

With this as a frame of reference, two system configurations have evolved: (1) the legacy/integrated enterprise systems view and (2) the best-of-breed view—an analytic application view. These views are distinct and dependent on the orientation and knowledge base of the viewers.

INTEGRATED ENTERPRISE SYSTEM VIEW

System suppliers and ERP vendors have announced that they are developing or integrating ABC into their enterprise offerings. Being very strong integrators of suites of tools, they have recognized the great need for ABC/M methodology in their customers and have chosen to integrate this tool set into their offerings. Although these players have dominated the transaction processing world, they are now entering a new world of periodic and analytic ABC/M analysis. Here the demand and knowledge expectations are different, and they are responding with their own perspectives. Some may come at ABC/M through performance management, while others may enter the market from manufacturing and operations perspectives. All in all, this is great for ABC/M as it legitimizes the business of ABC/M. Integration is of great value to customers as many of the problems of making ABIS and ABC work in pilot programs have been related to data translation and cleansing. It is hoped that the promise of ease of data accessibility will assist in more ABC/M implementations and successes.

BEST OF BREED: ANALYTIC APPLICATIONS

Besides the world of transaction processing and ERP, another strong component of the ABC/M landscape has been comparatively small software companies that specialize in ABC/M niches. International Data Corporation (IDC) has coined the term "analytic application" to identify this growing market.[8] All firms in the market must conform to a few criteria[9]:

- Process support: Structured approach to automating a group of tasks with relationship to business operations or in the discovery of a new business opportunity
- Separation of function: Functioning independently from core transactional systems but still using information from these systems or returning information to them
- Time-oriented, integrated data: Bringing together data from multiple sources and supporting time-based dimension for trend analysis

Analytic applications serve a separate yet defineable space in the ABC/M landscape. Besides their analytic value, these systems are easier to implement, use, and model pilot programs. With minimal investments, Global 100 corporations have "piloted" their implementations while they waited for the operational ERP systems to be implemented.

Analytical applications brought together can form a best-of-breed implementation for many firms. Those that have employed these methods tend to believe that one vendor cannot be best in every tool set. Certain Global 100 companies do not rely on one vendor of choice, controlling the integration of these best-of-breed tools into a standards-based integration while picking the best software and technologies in each class. Since the interconnectivity standards are establishing themselves in the software world, best-of-breed software systems can be combined with relative ease.

THE CHOICE IS TO DO BOTH

In the past, the choice of information technology depended on the footprint—the nature and expectation for an implementation. Each choice has its advantages and disadvantages; the only wrong decision is not making one.

Contemporary theories dictate that hybrid choices are made; both options 1 and 2 may be employed because it may take time to implement an ERP solution and ABC is needed now.

Furthermore, Kaplan and Cooper have offered an interesting approach to activity-based information systems. They state that installations go through several "stages" of evolution, namely:

- Stage I: Broken implementations of financial systems
- Stage II: Financial reporting driven
- Stage III: Stand-alone ABC systems with shared databases
- Stage IV: Fully integrated ABM and performance measurement systems

They clarify that companies that migrate from Stage I to IV without passing through the experimental and learning stages of II and III are destined to fail and repeat Stage II over again.[10]

Once its implementers have learned the practical tips involved with ABC/M in the Stage III systems, they can move to a full-blown activity-based information system that is tightly coupled with the real financial backbone of the company. Interpreting Cooper and Kaplan, Stage IV is realized when ERP systems are implemented with integrated ABC/M modules that run fully integrated.

ANALYTIC ABC APPLICATIONS REMAIN POST-ERP IMPLEMENTATION

When transaction-based systems incorporate ABC/M, much of production-level ABC/M should be easier than it is today. ERP systems, by their nature, tend to be directed to efficiency and operational effectiveness. Analytic ABC/M systems are the prototyping, learning, and training systems prior to an ERP entrance into the ABC/M project. (See Exhibit 6.1.) Post-ERP involvement in enterprise deployment, these analytic systems serve the modeling and prototyping need even further as new information drives new experimentation. In the organizational system architecture, such systems must exist off-line and be connected for such predictive and experimental purposes. For example, the ERP systems may pour information onto such a system for "off-line" experimentation and modeling.

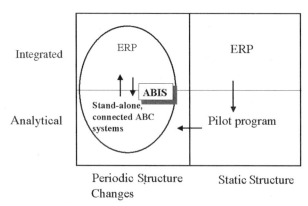

Exhibit 6.1   ERP-Analytical Matrix

WHAT IS ENTERPRISE-WIDE?

The ultimate goal of any system and methodology implementation is an enterprise-wide implementation of ABC/M. Single PC-oriented application suites cannot claim to be enterprise-wide in nature and scope unless tight links exist between them and ERP systems. Some analytic applications have claimed to be stand-alone and disconnected yet serve the enterprise in nature. This is highly premature. What they are doing is testing enterprise ABC/M prior to a true virtual enterprise system deployment.

Eventually PC-based information systems will be deployed throughout the world and linked with ERP systems via sophisticated technology.[11] In order for this to occur, a true partnership must exist among system vendors as a hands-off approach among them may not serve the client.

COMPONENTS OF BUSINESS INTELLIGENCE SOFTWARE SYSTEMS

There are three main components to business intelligence software systems. (See Exhibit 6.2.) They are:

1. Data warehousing technology
2. Data navigation technology
3. Business modeling technology

These technologies are dramatically improving the gathering, analysis, and deployment of information. However, the true power of BI has yet to be real-

COMPONENTS OF BUSINESS INTELLIGENCE SOFTWARE SYSTEMS 61

Exhibit 6.2   ABBI System

ized. Exhibit 6.3 illustrates how data warehousing technology combined with strong business modeling tools and desktop power can accelerate relevant information onto the desktop for quick and clear decision making.

Data Warehousing Technology

Data have always been stored in systems. The developing dream is to have relevant information any time, anywhere, and on demand. Data warehouses are not merely corporate data stores in the minds of most users. They are the sources of most intelligent information. Data warehousing was probably selected in the organization as a strategic imperative for:

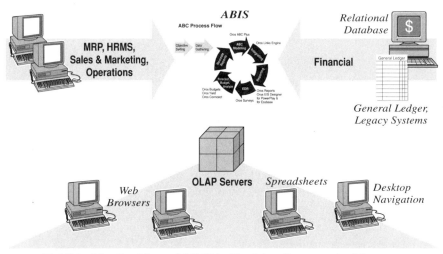

Exhibit 6.3   Position of ABIS in Decision Support Architecture

- Standardization in technology
- Strategic control and deployment of data to reach the desktop
- Consistency in reporting frequencies
- Consistent storage and retrieval of information

Data warehousing has been mistaken for one data storage device or central repository. Issues of redundancy, data integrity, security, fault tolerance, and performance will face anyone involved with it. Data warehousing is not a product or device; it is a strategy first of all. Essentially, data warehousing is a conceptual model for storing information for retrieval, and it could employ several physical systems and devices.

Unfortunately, data warehousing has been identified as a solution for any information retrieval system. With this view, the systems can turn into giant data trashcans. Information, as mentioned before, can age and atrophy. If data warehousing is treated like trashcans, organizations may find that when they go through the data items, they may not be able to trust the timeliness of information displayed. Many times, data warehouses are just not enough. On one hand, organizations need consolidated and organized data. On the other hand, timely unconsolidated, and nonhistoric, data specific to a user's need may just do the trick. These two demands have forced information technology teams to "augment their data warehouses with repositories that store production data in all its detailed glory."[12] If treated like a closet, information may be thrown into data warehouses with no regard to what it is related to; unrelated information is only data. Data must be structured to be meaningful. Data warehousing clearly is one part of a strategy in serving the needs of an enterprise. If combined with critical technologies, data warehousing can be a primary partner to the business intelligence architecture.

Data Navigation Systems

Data warehouses are generally "relational" database management systems (RDMS), which were invented by E. F. Codd to help manage transactions. The best way to describe a relational database is to view it like a file cabinet with files in it. Each piece of paper is a relational data element. On-line transaction processing (OLTP) systems, which is what relational technology builds, have populated the world as the de facto data stores, but they were never designed for rapid, multidimensional analysis. Consider doing rapid and multilinked analysis using a file cabinet filled with paper. Doing so would take time and tremendous cross-referencing.

OLAP (on-line analytical processing), another invention of Codd's, allows

COMPONENTS OF BUSINESS INTELLIGENCE SOFTWARE SYSTEMS

for rapid, multidimensional analysis. The demand placed on systems is increasing as more and more data that are unrelated is stored for relationships to be generated and understood. OLAP is like a Rubik's Cube. Several dimensions can be viewed. Information can be related to each other rapidly and beyond two dimensions. These systems also are called multidimensional database management systems (MDBMSs). Many of these systems have roots in decision support systems (DSSs) and executive information systems (EISs) of the past.

Peter Ruber from *VarBusiness Magazine* clarifies the importance of multidimensional analysis when he declares: "Multi-dimensional analysis is a requirement in organizations in which users must share and analyze complex, interrelated numerical data in areas such as product profitability, yield, sales performance, budgets, strategic plans and forecasts."[13]

OLAP has been viewed as the rapid-fire connection to OLTP systems. New hybrid systems have emerged as multidimensional OLAP (MOLAP) and relational OLAP (ROLAP). Exhibit 6.4 describes OLTP and OLAP. Note that MOLAP and ROLAP are two different but effective ways of viewing OLAP architectures. Fundamentally, OLAP frees information locked up in the "data jailhouses" found in mainframe databases that are OLTP in nature.[14] ROLAP does not store data; it is actually a set of data extraction tools that grab data from relational databases, that is, ROLAP is a mapping on top of a relational database. MOLAP stores consolidated or "rolled-up" data just below the individual transactions. The advantage to ROLAP is that maintenance of another database besides RDBMS is not required.[15] The maintenance burden is lowered.

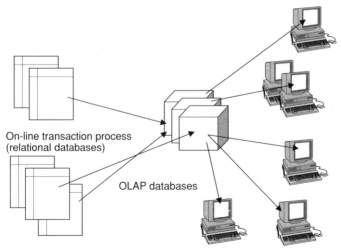

Exhibit 6.4   OLTP vs. OLAP

Furthermore, OLAP engines can be server based or client based. Some software vendors have taken OLAP to the desktop as well as the server. These systems have provided a new paradigm for users:

- Navigability: The ability to traverse from one dimension to another while drilling down in cause-effect isolation
- Multidimensionality: The ability to change dimensions to deal with greater than two reference points and the ability to relate information not usually connected (e.g., list all activities that are profitable and non–value-added)
- Hierarchical information: The ability to deal with multiple levels of abstraction in an analysis.

Exhibit 6.5a and b gives an example of the power of OLAP in drill-down analysis. Note that traditional unidimensional "drilling" will lead to the wrong conclusion. Merely by adding one dimension more of information, the questions turn to new and more knowledgeable answers. Unidimensionally, the wrong decisions would have been made. Consider, then, the power of multidimensional analysis. Initially these tools were developed for departments to view their own data, that is, sales and marketing to view their own measures or for finance to view their own proposals.[16] OLAP is enjoying rapid usage in cross-functional information in the ABC/M market. Now, users in one department are viewing consolidated information about various other departments.

Business Modeling

The momentum generated by OLAP and data warehousing is staggering. These technologies are feeding the deep need of corporations for information to help them understand themselves and to predict and model their future. Thus far predictive tools such as these lack structure and a conceptual model to frame the information displayed. ABC is one of a few conceptual frameworks available to model a business. Being cost-centric, most ABI systems cannot model all aspects of a business. However, these systems are moving to incorporate more comprehensive views of activities, revenue, cost, process, and performance measurement. As the maturing technology, activity information, and the need to predict and model the business converge, the field of activity-based business intelligence will surface as a powerful business modeling solution with strong underlying and supporting technologies. A leading authority on data warehousing, Shaku Atre, states: "What's needed and what vendors are starting to supply is depth of business content that lets you exploit the data

COMPONENTS OF BUSINESS INTELLIGENCE SOFTWARE SYSTEMS

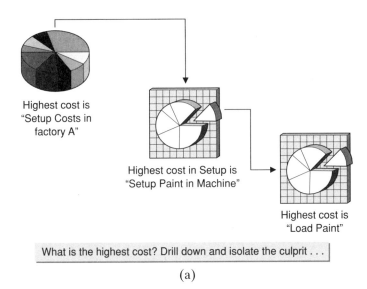

(a)

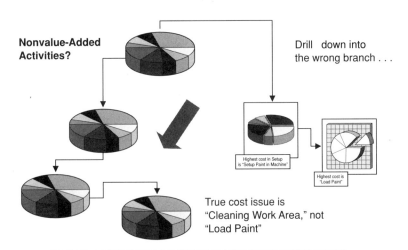

(b)

Exhibit 6.5 (a) Traditional Unidimensional "Drill Down" and (b) Multidimensional Drill Down Gets to Real Root Causes

warehousing opportunities for the industry (vertical applications) or business functions (horizontal application). This content needs to be built-in."[17]
Today, organizations can:

- Model their enterprise using ABIS
- Pour relevant related and dimensional information into an OLAP engine
- Program the structure and information into the cube
- Provide from the OLAP engine a connection onto the desktop where executives and users can navigate relevant activity-based information regularly for analysis

In an *activity-based intelligence system,* a person can accurately analyze financial information, relate it to the actual activities a company performs via such a system, and perform multidimensional analysis "drilling-down" to uncover causes and effects.

NOTES

1. Robin McNeill Connelly, and Roland Mosimann, *The Multidimensional Manager: 24 Ways to Impact Your Bottom Line in 90 Days* (Cognos, Inc., 1996).
2. Id., p. 3
3. Id., p. 92.
4. Steven M. Hronec, *Vital Signs* (Arthur Andersen & Co. 1993).
5. Robert S. Kaplan, and Robin Cooper, *Cost & Effect* (Boston: Harvard Business School Press, 1998).
6. Steve Player, and David Keys, *Activity-Based Management—Arthur Andersen's Lessons from the ABM Battlefield* (New York: Mastermedia Ltd., 1995), p. 10.
7. Robin Cooper, and Robert S. Kaplan, "The Promise and Peril of Integrated Cost Systems," *Harvard Business Review* (July–August 1998), p. 110.
8. Henry Morris, "ABC Technologies: A Business Methodology Foundation for Analytic Applications," *IDC Bulletin* No. 16838 (August 1998).
9. Henry Morris, "Packaging the Vertical Warehouse: Cognos' Application Partnerships," An International Data Corp. White Paper, November 1996. *IDC Bulletin,* No. 18340.
10. Kaplan and Cooper, *Cost & Effect.*
11. Russell Shaw, "ABC and ERP: Partners at Last?" *Management Accounting* (November 1998), pp. 56–58.

NOTES

12. Craig Stedman, "Augmenting Data Depots," *ComputerWorld,* June 16, 1997, pp. 57–60.
13. Peter Ruber, "EIS: Selling to the Top," *VarBusiness* (October 1994), p. 92.
14. Aaron Zones, "Building a Business Case," *Communications Week,* August 29, 1994. "More and more, business managers—from corporate executives to departmental and project leads—require reliable information to derive day-to-day and long-term business plans. Yet, much of the data that would provide the foundation for such analytical processing remains locked up in 'data jailhouses' mainframe databases that are optimized for on-line transaction processing."
15. Stewart Mckie, "The Power of OLAP," *Controller Magazine* (January 1997), p. 35.
16. John Foley, "OLAP Spread," *Industry Week,* October 20, 1997, pp. 20–22.
17. Shaku Atre, "From Build to Buy," *ComputerWorld,* February 9, 1998, p. 66.

7

SEVENFOLD WAY: IMPLEMENTING ACTIVITY-BASED INFORMATION SYSTEMS

Today many information systems deploying desktop navigation and data warehousing strategies using OLAP engines are efficiently speeding irrelevant information on the desktop in lightning speed. These systems have employed the technologies of business intelligence but not the intent behind it: to deploy and receive relevant related information in the enterprise that reflects the enterprise in a complete fashion. Say a manager wishes to reduce the cycle time. She will look for strategies to do just that. What if she learns that cycle time can be reduced by 20 percent or by 50 percent? The obvious choice is 50 percent. By looking at the activity costs of these choices, she discovers that 50 percent improvement will double costs while 20 percent will merely increase costs slightly. Her choices have changed due to the addition of related and relevant information. Moreover, with activity-based analysis, organizations can learn what not to do in their factory.

Relationships between elements of data create knowledge; activity-based relationships create activity-based knowledge. Thus far transactions and their behavior have been the mirrors of organizations. This is called *transaction-based business intelligence,* as the general ledger is used to conduct business. This information is very department-centric and department-used. Sales personnel would input sales information and observe this information in aggregate while they perform ad hoc analysis. Sears, Roebuck and Co. discovered that since the early 1990s, their merchandisers looked at merchandizing information while their retailers' financial personnel looked at financial data.[1] As the business challenges matured, there seemed increasing need to view each other's data as well.

Systems now allow for integrating transaction information with the general ledger using financial reporting tools. Seemingly related information can be viewed and analyzed. Still, using all the drill-down analytical tools, the analyst

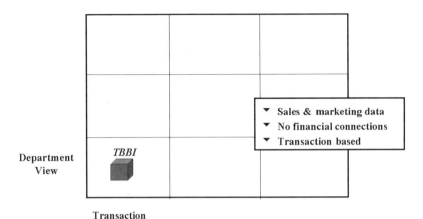

Exhibit 7.1 ABBI Evolution

ends up in the general ledger items, which does not provide the operational teams with what actions they can perform. This is *ledger-based business intelligence*. While this is valuable information, department views exist and they reflect departments and structures that are not process or activity-centric. Operating teams continue to analyze their behavior through the lens of the general ledger.

Both these views of information cannot show enough of the intricacies of work being performed and the impacts of work to the organization. They also do not take a process-view of the organization. Exhibits 7.1 to Exhibit 7.5 illustrate the progression of information understanding and deployment.

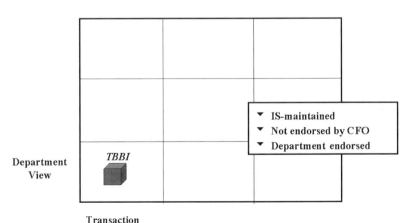

Exhibit 7.2 ABBI Evolutions

70 SEVENFOLD WAY: IMPLEMENTING ACTIVITY-BASED INFORMATION SYSTEMS

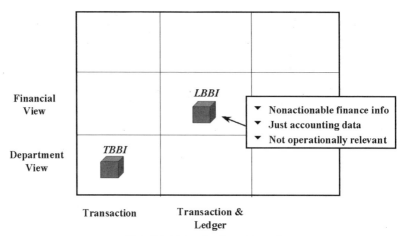

Exhibit 7.3 Ledger Based

Information like this is seldom used operationally, even though data come from finance and sophisticated information systems are used to gather and deploy them. Information that is operationally used, finance-endorsed, and IT-maintained is information that translates financial performance metrics into the language of operations—into activities. If activity information is available, operating teams can now trust and act on financially endorsed information that comes to them regularly through the IT systems. Furthermore, the information reflects the work they perform—activity-based work. Activity-based information relates the resources given with the work to be performed and the products and services created from this work. True business intelligence can be

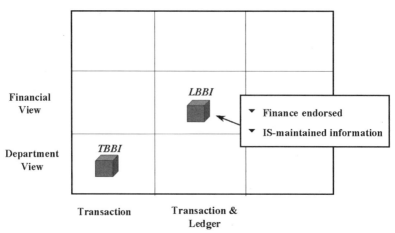

Exhibit 7.4 Who Supports LBBI

SEVENFOLD WAY: IMPLEMENTING ACTIVITY-BASED INFORMATION SYSTEMS 71

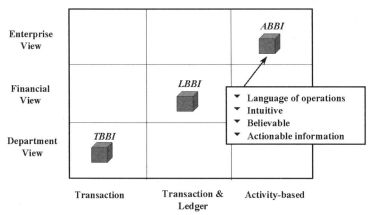

Exhibit 7.5 ABBI Evolution

achieved with activity-based analysis and technology. ABIS is enabling technology while ABC/M is the information that is produced from the use of methodology and the ABIS systems. ABBI is the regularly maintained, finance-endorsed, operationally used information that operational teams used to take action. With these fundamental components, ABBI (activity-based business intelligence) is a reality.

ABC/M projects, in isolation, can be viewed as futile activities if they are run with no visible structure and procedures. If not focused and mission oriented, ABC/M projects can be useless. The sevenfold way illustrates a method that combines the success conditions identified by several organizations. This method is not a recipe to success but a collection of conditions found in several successful ABC/M implementations. However, each did employ parts of this approach and saw results.

Exhibit 7.6 illustrates the sevenfold way. Its premise is "knowing before doing; doing before showing." Note also that the premise contends that organizations have little control over the outcome of a project; they may have control over establishing the preconditions of a successful implementation.

- The First Way: Understand the Organization
  Often organizations get into the ABC/M fever prior to understanding their own capability and capacity to do ABC. Usually costs are not controlled or the causes of costs are elusive in organizations before they embark on ABC. Organizations should know themselves well enough to know what they are willing to do once armed with ABC information. Organizations have biases and have ways to solve problems. ABC/M can succeed only if it adapts to these biases and perspectives.

72 SEVENFOLD WAY: IMPLEMENTING ACTIVITY-BASED INFORMATION SYSTEMS

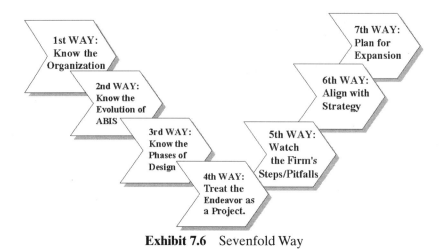

Exhibit 7.6   Sevenfold Way

- The Second Way: Understand the evolution of activity-based information systems.
  In this fast-paced evolutionary curve of new technology, it is best to know where organizations are on this curve. ABC/M projects go through phases of learning and development. Activity-based information systems have also evolved. Understanding how they evolved in the phases will help organizations pace themselves and watch for the signs that permit them to move along in the phases.
- The Third Way: Know the phases of designing an ABIS.
  Activity-based information systems take careful time and energy to use and deploy. Like any other information system implementation, any ABIS work requires a phased approach, from objective setting, to designing the models, to reporting.
- The Fourth Way: Treat the endeavor like a project and the results like a product.
  Several successful ABC/M implementations have learned that the best way to make ABC/M a part of the business culture is to run the ABC/M program using the same project management methods employed in the organization. If organizations are product and project oriented, making the project have ship dates, product freeze schedules and product life cycles only enhances the chances for use. Furthermore, treating the clients of the information as your customers forces a formalism for success.
- The Fifth Way: Watch the firm's steps, beware the eight pitfalls.
  Experience has uncovered eight major technical hurdles that reduce the chance of success in an ABC/M project exercise. These hurdles seem sim-

ple when identified upfront but are dangerous when unanticipated and can cause many a project to fail.
- The Sixth Way: Take the first step and align with strategy.
  The number one reason why ABC/M projects fail is lack of upper management support. Unfortunately, those who hold the purse pay the bills. ABC/M projects are initially based on trust; that is, people do not really know what they will get but know the value of the project only after it is completed. Organizations want do ABC/M because they are in trouble of some sort in the first place. Their patience is not high from the start, and they are searching for the magic pill to solve all ills. For an ABC/M project to come through, the executives must believe in it and what it can produce. Strategy is the weapon and language of upper management. Cost is the language of finance. Both must meet if ABC/M is to survive in any organization.
- The Seventh Way: Expect and plan for expansion.
  Many organizations are so excited about pilot programs that they charge like rhinos in a straight line and at high speed. Their passion drives them to work long and hard hours demanding everything from everyone. They gather data through interviews and use hand-drawn process charts to accelerate the results. They finish the project and produce the finest reports in the history of the company. The project lead gets a large and heavy plaque for his or her cubicle and the celebration begins. The next morning the chief executive visits the champion, asks that the project be performed across the entire company, and promotes the champion to director of activity-based management. But the chief executive wants the results faster than before. In fact, he wants reports monthly to analyze all 20 manufacturing locations starting in three months. If the ABC/M champion did not plan for enterprise-wide modeling and deployment, this project would take five steps back before taking any steps forward. Certain elements in planning can reduce this challenge.

The next seven chapters outline, in detail, each of the seven ways and how they assist in successful projects.

NOTES

1. Thomas Hoffman, "Datawarehouse, the Sequel," *ComputerWorld,* June 2, 1997, pp. 69–72.

8

FIRST WAY: FIND THE FOOTPRINT

Organizations frequently dive into ABC projects like eager children jumping into a swimming pool on a hot summer day. Yet the enthusiasm is dwarfed by the utter lack of understanding some organizations have in the venture they embark upon.

The challenge many ABC teams take on is perceived as a change management challenge. Many of these teams are led by dynamic professionals who are determined to show a better way to their organizations. They are progressive, rule-changing, and intelligent change agents who see possibilities of change but may not be accustomed to its politics. Many are rationale-centered people—they believe that logic and results speak for themselves; that truth will be understood and even embraced by management.

As Steve Player of Arthur Andersen says, "Generally speaking, wet babies are the only people who enjoy change and yet change is constant in all entities."

Generally, finance professionals have taken advisory roles in major corporations. Their primary functions were to provide useful analysis and reports to management whom, they believed, understood the issues and the critical decisions to be made. Now finance professionals are being transformed from reporting engines to business partners. PricewaterhouseCoopers LLP, in a major 1989 study, uncovered that 75 to 80 percent of chief financial officers' time was spent on fiduciary issues—outside reporting. In the coming century, it is expected that this figure would reverse, with CFOs spending only 10 to 20 percent of their time externally.[1] CFOs would spend the rest of their time internally, partnering with operations in decision making and consultation. This may lead us to believe that finance professionals must upgrade their change management skills to win with ABC/M. That is only partially true; finance professionals are more challenged to learn more about the business of operations. "There's no reason so many people are cynical of canned change

FIRST WAY: FIND THE FOOTPRINT

programs and distrustful of the 'change weenies' sent to administer them. It's the same reason so many change programs fail: They have nothing to do with what really matters in business."[2] Clearly, it is the task of change leaders to show relevance and also to be business leaders before being change leaders.

One Fortune 100 company recently conducted an ABC/M project with exactness and enthusiasm. The CFO was bought into the process totally; the technology development was on target and almost overengineered for all probable questions and answers. The company needed the information because the trigger factors of survival and profitability existed. The company tried several times to educate the operating teams, but they could not seem to embrace the concept of ABC/M and recognize its value.

The operating teams agreed that new product and service pricing was valuable but could not believe that the numbers they were receiving from the general ledger were wrong. How could information they had been receiving for so long be incorrect? Age-old studies have proven that collective representations of reality are self-fulfilling prophecies—they confirm each other's thinking because the basis of their belief system is the same.[3] In this sense, denial loves company. This Fortune 100 firm performed exactly as expected; it denied the information was true because more than one person said so.

The major aspect of this illustration is to recognize that no matter how valuable the information and experience, human perception can hold back an ABC/M project. And reality is still in the eyes of the beholder in business today. If people do not understand the way in which their company deals with truth and information, they will never get anywhere with new information. Many companies that have failed in several initiatives tend to believe that the next one will do the job. This naive notion is the real reason why they fail—they lack an understanding of their incapacities and skills. Other companies succeed in their ABC/M initiatives because they have a strict culture for change and can make extensive information system changes without negative feedback.

Managers at Saturn Corporation understood their organization enough to claim that "a man with two watches can't tell you what time it is."[4] This observation may lead you to think that all general ledger–based reports would be removed once any ABC/M-based reports were on-line and active. Managers at Saturn wanted to remove fiefdoms of information, and they did this by forming one common database. Saturn's secret weapon seemed to be its philosophy rather than its technology. The company's primary finance and accounting philosophy is that the finance team has an obligation to train everyone on the financial aspects of the business.

What was the real reason behind the Fortune 100 company's lack of change? Why was it so clearly resistant to ABC/M when everyone around them seemed to understand and work with the information? The breakthrough happened when one of the consultants was approached by an executive after an internal ABC forum. This executive told the consultant that he wanted to learn how to ask the right questions and make decisions—in other words, to think multidimensionally.

The complexity of modern organizations has overgrown the skills of many operational managers who still think unidimensionally. The ability to think and ask questions multidimensionally needs training and education. ABC/M projects are becoming 10- to 20-dimensional models. Databases gave us answers to basic one or two-dimensional information like "what is the cost of x?" Current technology allows us the ability to ask "What is the profit of all products, by the channel, by the regions they are sold, in ascending order, and the activities that feed them?" If we do not know how to ask the right questions, how can we get to the right answers? A technology manager at an international restaurant chain noted that her staff spent the first six months after implementing a data warehouse responding to the users of the information.[5] Many organizations that embark on the ABC/M journey do not understand what they are creating and whether their organizations are willing *and* ready for this challenge in rethinking. When embarking on its data warehouse project, MCI Communications Corp. alleviated this challenge somewhat by introducing sample versions with subsets of information to build "pockets of knowledge and familiarity."

FROM PITFALLS TO ENABLERS

As mentioned, over 30 pitfalls have been identified in the implementation of an ABC/M project,[6] based on the experiences of several early adopting organizations. Merely avoiding them would get users far. But let us look at the conditions that create success rather than the conditions to be avoided.

Once Andy Grove, head of Intel Corporation, illustrated the value of planning by asking for the best decision among three alternatives in the following situation: A person has run out of gas along a highway. The obvious choices can be:

- Walk to the nearest gas station.
- Get a ride from another driver.
- Call for assistance.

Each alternative has pros and cons, but the point Grove made stayed with me for years. He asked why the driver ran out of gas in the first place.

In ABC/M projects, fuel is critical to the trajectory of the program. How the project moves from pilot to production and then to full deployment depends on the fuel generated and carefully deployed throughout the journey. Prior to beginning any journey, people must measure and observe their capacity to take the journey without falling prey to the pitfalls. Otherwise they will spend more time getting out of the pitfalls than creating solutions. Many times understanding the people and their organization's readiness to change determines the fuel, which is a key quotient for success. This can be called determining "task-relevant readiness."[7]

The readiness and context maturity is the organization's ability to respond to and strive for outputs from an ABC/M project. On occasion, I have the difficult task of counseling program managers who are about to launch their ABC projects. Almost all ABC-related conferences can educate you on what to avoid and how to make things great. Yet two main ingredients to ignition tend not to be addressed:

1. Is the champion ready to lead?   Task-relevant leadership quotient
2. Is the organization ready to follow?   Task-relevant readiness quotient

Task-Relevant Leadership

Is the champion ready to lead an ABC/M program? Can the champion absorb this great responsibility and ensure that it succeeds?

ABC/M pilot projects benefit a great deal from targeted, passionate leadership. Leaders of the projects often have a mission—to improve and change the enterprise. Anyone who leads or is about to lead an ABC/M project probably has a strong desire to change the way the business around him or her behaves. Generally, the typical ABC project is profiled in the following way.

- The champion reports to the CFO/finance managing director of an organization.
- The project is finance initiated but the goal is operational usage of information for decision making.
- It is steering committee driven. Most projects have an executive committee overseeing the program's activities.
- Powerful executive sponsorship exists.
- Most projects also use a cross-functional team to drive the program.
- The project uses pilot off-the-shelf but connected software.
- Almost 50 percent of the time, an external consultant is used. Other times internal consultants are engaged.
- Most teams are charged with a set of objectives.

- Most of these teams are looking for the "aha" effect, that is, looking to be surprised by their discovery instead of merely establishing a methodology in their organization.

Certain definite traits of successful ABC/M program leaders surface in the course of an ABC/M exercise. Consequently, many ABC/M leaders tend to:

- Be renegades and nonconformists
- Enjoy "missionary" selling and are convincing in their communication skills
- Enjoy and demand change
- Are impatient with the way things are
- Are rapid learners and enjoy knowledge above routine actions
- Are ABC/M knowledgeable
- Somewhat cynical about the way business is performed today
- Feel lonely in their mission and sometimes feel that very few understand their goals
- Wonder if ABC/M will ever grow in their industry
- Have an eye for detail but have the big picture in their view
- Enjoy being visible to upper management
- Be frustrated about limited resources on the project because the mission is above any other goal
- Believe that they need more support from management and that management ought to just order everyone to follow because it is so obvious
- Have a strong entrepreneurial spirit and associated skills to build a small company within a big one
- Believe that they cannot last long in the ABC/M program and will have to move on after
- Be skilled communicators with IT groups, finance teams, and operational teams
- Enjoy getting past "bean counting" to work with the operational teams
- Motivate volunteers or part-time assignees to the program
- Get bored easily and like movement
- Enjoy the rush of the new challenge
- Not to establish generally acceptable standards
- Not to wait for routine procedures and policies but like to get to the right answers
- Not look for repeatability, scalability, and maintainability
- Get things done without asking permission but asking for forgiveness

Each team member may not have all these skills and traits; team members should share them.

Task-Relevant Readiness: Is the Organization Ready for ABC/M?

ABC/M usually begins as an initiative within most organizations. These firms want to analyze and change the way they view their business, activities, products, services or channels.

The most documented reason why ABC/M projects fail is "the lack of upper management support." Nine out of ten case studies mention this phenomenon. Although the best way to improve the chances of implementing ABC/M is to get management support, this borrowed authority and endorsement is only a smokescreen to a more systemic problem—the readiness of an organization to accept, embrace, and use new ideas and concepts for improvement.

An organization that has failed to implement TQM, business process reengineering, or process improvement is probably highly likely to fail in implementing ABC/M. Any of these change programs may have had management support and guidance at some point, but they still failed. ABC/M has the added support of information systems, which help in sustaining the initiative. There are four aspects to gauging and building the organization's technology readiness for ABC/M.

PREPARATION FOR THE ABC JOURNEY

Before embarking on an ABC journey, users should try to understand and then plan through four aspects of readiness. (See Exhibit 8.1.) They are:

1. Collect the key ingredients to launch implementation.
2. Align the the program to the organizational personality.
3. Educate the enterprise.
4. Move from agreement to commitment.

Collect the Key Ingredients to Project Ignition

As CASE Corporation's GENESIS project has learned in the course of reengineering its finance group: "The pilot project had to demonstrate results to management, and the results needed to show not only how to get information, but also how to use information and benefit from it."[8]

Organizations usually resist change, even when change is the only way they can survive. Why should anyone be surprised that teams resist change? The true change agent is attracted to resistance. Like metal to magnets, ABC leaders enjoy the challenge of change.

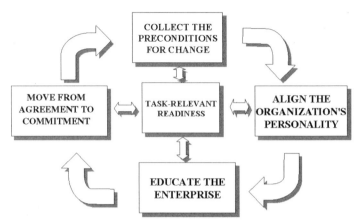

Exhibit 8.1 Four Aspects of Organizational Readiness

Organizations have similar yet contradicting profiles. They are:

- Change fatigued—too tired of change programs. Many just wish to wait it out until the chief executive officer gets another idea. Some organizations actually enjoy anticipating change programs by calculating the number of books the executive officers read.
- Questioning the business value—too frustrated with quick fixes. Many organizations will debate the value of any new methodology, especially when it means that resources have to be applied.
- Observing who sponsors the program—wary of politics, organizations will watch which way the wind blows and will tend to follow when they believe that there is no way out in the short term. Change agents should not mistake agreement for commitment in ABC/M programs. In order to work, ABC/M requires strong buy-in from key change sponsors. The rest of the organization will be transformed by observing their commitment.
- Focused on measurement and rewards—"Measure me and I will move. Reward me and I will leap" seems to be today's business battle cry.

These are the basic lowest common denominator conditions in many of today's organizations. Change agents who wish to perform a successful ABC/M exercise must understand and work with these conditions. These perspectives replace fear, uncertainty, and anxiety with momentum and commitment. Projects can go in one of two ways: They can ignite or they can combust. Project ignition demands that the ABC/M champion bring the following preconditions into focus:

- *Vision.* The ability to see and articulate the way things can be
- *Knowledge.* Awareness built on keen, leading-edge thinking and analysis
- *Experience.* Practical, hands-on awareness derived from "being there, doing it"

If applied to the three main enablers in any ABC/M project—people, process, and technology—all three of these preconditions can reduce risks and increase the probability of success. (See Exhibit 8.2.)

ABC champions must identify the sources of these preconditions and apply all these available resources to the project.

Align the Organizational Personality to the Project

At Dayton Extruded Plastics, ABC implementers encountered their first problem when they interviewed marketing and sales before the team truly understood the value of ABC/M. The implementers heard early criticisms on the project. "This brought out some wonderful discussions about overhead, margins, profitability by parts, and other thoughts about how this system can be used to estimate profits and costs on parts as they are being developed," an implementer states. "With these, lively discussions turned sales and marketing into eager participants."[9]

What Dayton learned, organizations embarking on ABC/M should plan for. How does the organization adopt new ideas and processes? What filters are

	Vision	Knowledge	Experience
People	Executives, CEOs	Educators, thought leaders	Practitioners, ABC modelers, project leaders
Process	ABC/M life cycle, project management, objectives setting, and decision-making expectations	Educators in the industry, consultants, modelers	Industry-specific examples, cases, best practices, worst practices
Technology	Chief information officers, ABC committee	Technology consultants, software vendors	Software vendors, IT department resources, user group forums

Exhibit 8.2 Relationship of Preconditions to Enablers

used to understand, adopt, or reject new technology implementations and new endeavors like ABC/M?

Organizations tend to have one of three focuses: (1) people, (2) process, or (3) technology. Understanding the biases of an organization is critical to the successful implementation of an ABC/M project.

For example, introducing new technologies early on to a primarily process-driven organization can be a mistake. Conversely, ABC champions who talk about people issues to a primarily technology-focused organization may not get results. Understanding the organization's bias and viewpoint will guide implementers in introducing the ideas and technology of ABC/M. While ABC champions usually are very focused on introducing the concept of ABC to their peers and management, they will be more successful if they first understand the way and through which lens these peers and management view the organization. Some stereotypical perspectives for all three views follow.

People-focused organizations tend to:

- View their world through human issues.
- Believe that if people are motivated and happy, all is well.
- Believe that profits are important but people must be content for profits to be achieved.
- Believe that layoffs are traumatic and not an option.
- Believe that firings are contemplated for a long time.
- Train their people and develop them.
- Permit human resources and management to guide the company.
- Believe that communication is key.

Process-focused organizations:

- Are really excited about organized initiatives.
- Get things done by project management.
- Are TQM, process-centric.
- View the world through finite processes, activities, and tasks.
- Value people who are members of a process.
- Tend to have operating teams rule.
- Are places where product life cycles drive products rather than products driving life cycles.

Technology-driven organizations are:

- Bits-and-bytes oriented.
- The first to upgrade systems.

- IT focused and motivated.
- Users of the latest and greatest technological development are early adopters of new technology.
- Very change driven and sometimes forget about evolutionary change and compatibility.

Naturally, organizations can display any one of these three biases at different times, and large organizations also can display various biases at the same time. Every organization has an underlying personality, which is probably not just one of these three biases but somewhere in between. In understanding the organizational footprint, consider where they fit in Exhibit 8.3.

In the exhibit, the organization deals with issues in this priority: technology, people, and process.

Taking a strong look at the ABC/M objectives and the ways in which to approach (or the organization has approached) a project, outline the personality and *center of gravity* of the ABC/M program. The center of gravity is the unique balance of the three priorities within the organization. If the ABC/M project is not aligned, then the project creates a paradox. The conflict arises in introducing the goals and objectives of the project which seems against the natural biases of the audience. Exhibit 8.4 shows such a paradox.

Exhibit 8.5 illustrates eight bias maps for project introduction and planning. When ideas and new methods are introduced to an organization, match their biases to find anchors for the project.

Educate the Enterprise

Those who have a plan to educate their teams and organizations tend to create and reinforce the need for activity-based information. ABC/M champions learn about ABC through books, conferences, and user groups, among other

Exhibit 8.3 Find the Organization's Center of Gravity

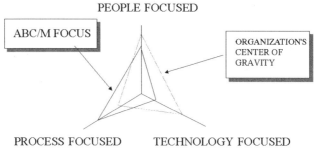

Exhibit 8.4 Find the Organization's Paradox

ways. Their learning is usually rapid while the learning of their organization is much slower. ABC/M project success can be corelated with the level of learning and commitment of the organization as a whole. For ABC/M to fully deploy knowledge to the desktop of operating teams and for them to make good decisions with this information, teams throughout the organization must learn about ABC and its value to decision making.

Chapter 4 discussed the technology adoption cycle in ABC/M as the following:

- Education
- Pilot
- Enterprise

Projects move from one phase to the next in sequence. (See Exhibit 4.1.) Individuals, teams, and organizations also take a similar learning path. In an ABC/M project, a number of types of inertia resist this learning:

- Satisfaction with the way things are
- Fear of the unknown; what is known is safer
- Hesitance to learn without tools and focus
- Too much to do
- Being measured on other performance measures
- Fear of loss of job security

Exhibit 8.6 illustrates the audiences that may require focused education in climbing this ABC power curve.

The seminal work of Peter Senge in *The Fifth Discipline* has focused us on the art and science of collective organizational learning. ABC/M programs start, grow, and die with "learning disabilities" as their poison.[10] Does the

PREPARATION FOR THE ABC JOURNEY

People	Process	Technology	Possible Condition
High	High	High	The project may lack focus. High achieving in nature.
High	High	Low	Focus on people and process issues. Structured methods are important. Technology patterns may be selected later. Technology training is necessary. Management may consider having consultants select technology.
High	Low	High	People and technology focused. Beware that human issues divert objectives. Fad technologies may divert project. Consider bringing in consultant/project manager.
High	Low	Low	People-centric. Insensitive to cause-effect relationships. Politics of people may kill project orientation. Priority may be to maintaining relationships.
Low	Low	Low	Nowhere land. Don't attempt anything.
Low	High	High	Process and technology focused. People/communication lacking emphasis. Consider consultant to contain project. Create communications systems. Remove fear.
Low	Low	High	Technology is worshipped. "Ready-fire-aim" mentality. Belief system: technology overcomes all. Consider process consultation. Consider communications consultant.
Low	High	Low	Process-centric. Acceptance criteria: fits into process. Technology training is necessary. Communication process may work.

Exhibit 8.5   ABC/M Bias Mapping Chart

Exhibit 8.6 ABC/M Power Curve

organization have any learning techniques that have worked in the past? How does it learn new ideas and implement them? Does it have blind spots—for example, does it not notice people issues? Exhibit 8.7 illustrates a sample, organized approach to identifying the learning tools in an ABC/M project.

UNDERSTANDING SELF DEMANDS CAREFUL OBSERVATION

Steven Covey often has been quoted as asking us to seek to understand before being understood.[11] Ultimately, the failure of the ABC/M educational process is seldom a failure of the team or organization. It is usually caused by a mismatch between the educational tools and techniques and the organization's internal methods of learning and communicating. Many organizations believe that to obtain this information, the CEO can dictate the needs and the methods of technology and methodology absorption. However, such an approach is relatively short-term. CEOs seldom keep on this topic indefinitely. ABC, if introduced effectively, never stays the same; it mutates to serve the way business is done rather than remaining an initiative. Before embarking on the journey, ask some of the following questions to gain a true understanding the characteristics of the audience. The questions themselves will wake up the audience and prepare them for what will occur next.

- What has the person heard about ABC/M in the organization?
- What are the objectives of this initiative?
  —To downsize?
  —To improve processes?
  —To discover true costs?
  —To understand overhead?

	Operations		ABC Team		Corporation		Chief Exec.	
	priority	phase	priority	phase	priority	phase	priority	phase
Presentations								
ABC conferences			9	e				
User group invites	1	p	2	p				
Videos			5	e,p				
Distance learning	2		7	p, pr				
On-site demos			6	e, p				
On-site "rapid prototyping"	3	e	3	e			7	
On-site seminars			4	e				
Public seminars	5	p						
Speakers							2	
Thought leader visits			1	e			1	
Newsletters from vendors	4	p						
Books	6	e						
White papers							3	
Success stories	7	p	8	pr			4	
Magazine articles								
E-mail messages							5	
Fax machine messages								
Templates of reports								
Web site/intranet info	8	pr					6	

Phases are:
e=education
p=pilot
pr=production

Exhibit 8.7 Resistance Busters: Tools for Educating the Enterprise

- Who sponsored the initiative?
- Are competitors using it?
- What is it?
- What are the key business issues we are seeking help to solve?
- Does the person think this is a Trojan horse for downsizing exercise?
- Have previous initiatives succeeded? Why? Why not?
- Does the person believe this is another initiative or a new way of doing business?
- How much time is needed for training?
- Has the person read or seen anything on ABC?
- What technologies are used, and how often?
- How much time is available to analyze data?
- What percentage of input is gotten from
  —E-mail?
  —Voice mail?
  —Internet?
  —In-basket?
  —Meetings?
  —Seminars?
  —Faxes?
  —Phone calls?
  —Forums?
- What is automatically removed from the person's desk?
- What is the biggest challenge ABC will face?
- How many people are involved with ABC?
- What is the standard database system used?
- Which data warehouse is used?
- How does the company budget?
- Which ERP system is being implemented and used?
- Does the IT group employ a best-of-breed tools strategy, or do people have to follow the specified systems?
- How often is the person surveyed?
- Should the firm hire outside consultants?
- Is the person part of a team? Does it follow a process?
- What is the optimal team size and composition?

ABC/M projects that take the time to analyze how information changes their clients' behavior can articulate reports that work. Otherwise decision makers receive reports but do not change behaviors because the reports do not reflect the interests and focus of these leaders.

INFORMATION CAN BEHAVE DIFFERENTLY

"Information behavior," which, loosely defined, means how people approach and handle information, introduces a very relevant tool to understanding the educational inabilities of an organization. Information behavior can be applied to how users use ABC information. Prior to developing a plan to educate the various constituents of the organization, information behavior must be understood as it affects how technologies can be planned, developed, and implemented.[12]

There are two ways to view how information is handled:

1. By shadowing users. Observation is far more powerful a research tool than questionnaires. Watch clients of the educational program and see what they read, watch, and learn.
2. By knowing their technology preferences: Watch what media they use to do their work. Do they read e-mail, intranet information bases, or use paper?

Try to map their specific preferences to the media available to educate and inform them of the ABC process.

HOW TO APPROACH EDUCATING THE ORGANIZATION

First, we must identify who the audiences are. In each phase, the audience increases to include new entrants, such as IT professionals, manufacturing personnel, and so on. Two strategies can be used to educate:

1. Rifle-shot approach
2. Buckshot approach

In the rifle-shot approach:

- People are educated only on a need-to-know basis.
- The project takes a skunk-works form, that is, a private and underground project, so that implementers can gain traction and prove the concept before others destroy the opportunity.
- Resistance may not come prematurely.
- Once proof of the concept is gotten, top management may not buy into it.

In the buckshot approach:

- Users are going for maximum leverage and breadth.
- The CEO and executive management are educated to gain consensus.

- Education is a major program in the company.
- All resistance must be addressed upfront while developing a proof of concept.
- Resistance will come in the form of debate.
- The project may never begin and be stagnant in debate.

MOVING FROM AGREEMENT TO COMMITMENT

Often teams agree that they will achieve a goal and try to do it to the best of their ability. Effort is seldom worth a passing grade when it comes to ABC/M projects. The Apollo 13 mission gained the world's attention recently when it was put on the big screen once again. The mission captured the hearts of Americans because it was filled with heroic deeds in the face of disaster. The best part was that the astronauts made it back alive against all odds. The drama and courage they displayed overshadowed the fact that they failed in their mission to land on the moon.

At the end of the day, there is no replacement for the commitment to completing your project with the necessary information to make the right decisions. For a project to succeed, the ABC/M project team and management must move from agreement to commitment by first knowing the difference between the two. Agreement is:

- We will try to make it work.
- It's good, we'll try it.
- We hope everyone sacrifices.
- If it fails, we know we would have learned a lot.

Commitment is:

- We will make it work.
- It's a sacrifice but the returns are clear.
- It will not fail because if it does, the corporation will fail.

The primary way to gain commitment for ABC/M is to get upper management buy-in to resource commitments upfront. Faheem Zuberi, director of finance and accounting for Sallie Mae, a leading provider of financial services for postsecondary education needs and the country's largest provider of funding for education loans, states: "Once senior management committed to the ABC pilot, the implementation went very well."[13]

Top management will buy in to ABC/M only if they recognize the true value that it can provide in realigning the operations of the organization. Many times

MOVING FROM AGREEMENT TO COMMITMENT

it is external forces that cause acceptance to ABC/M. One major pull for commitment is the change in market dynamics, such as consolidation in the banking market or loss of revenues and market share in the semiconductor business. In these situations, ABC/M may end up on the CEO's radar screen. If this is not the case, five common buy-in methods exist:

1. Taste-testing. In this method, using short and small pilots in an attempt to prove the concept and the opportunity may work. Using rapid prototyping techniques, the ABC champion can gain commitment and help the executive management visualize the solution.[14]
2. Mikey likes it! This technique involves using industry benchmarks to drive ABC/M as the method of choice. Within an industry, prove that a wave of ABC/M initiatives have produced results. Using the outside influence and need to drive internal acceptance will hasten commitment.
3. Doctor Says So! This technique involves using outside consultants and educators to convince management of the knowledge leadership the organization can attain. Books, videos, and other collaterals are used to convince management that these techniques have proven competent.
4. Trust the Wingperson. This technique involves using the informal leaders within the organization to influence management. Since they are trusted confidants, they may play the role of the trusted bishop whispering in the king's ear.
5. Ride the Wave. In any organization, many initiatives occupy the attention of management. Many of these initiatives may need to be measured and associated with costs. Here the aim is to convince management that ABC/M feeds or drives these key initiatives. In many organizations ABC/M has fed the performance measurement programs called key performance indicators (KPIs). KPIs are designed to establish and measure performance goals within teams. Teams are measured periodically on performance goals and are rewarded based on their achievement. Introducing ABC/M without any linkage to KPIs would be a recipe for failure. Many times there is tremendous linkage between these goals and ABC/M. Eileen Morrisey, head of ABM at Allied Signal, asserts that "As XYZ's managers reviewed their ABM cost drivers, they began to recognize that some drivers were ABC performance measures. For example, the number of customer complaints could be used both to measure performance and act as a cost-driver."[15]

Understanding the organizational footprint is the first step to designing a complete and successful ABC/M program. The second step is to understand how ABIS has evolved.

NOTES

1. Thomas Walther, Henry Johansson, John Dunleavy, and Elizabeth Hjelm, *Reinventing the CFO* (New York: McGraw-Hill, 1997), pp. 5–6.
2. Charles Fishman, "Change," *Fast Company* (April-May 1997), p. 66.
3. James O'Toole, *Leading Change* (San Francisco: Jossey-Bass, 1995), pp. 168–169.
4. Bala Balachandran, "Cost Management at Saturn: A Case Study," *Business Week Executive Briefing Services,* Vol. 5 1994, p. 27.
5. Craig Stedmand, "Do Users Know Your Data?" *ComputerWorld,* August 11, 1997, p. 37.
6. Steve Player and David Keys, *Activity-Based Management—Arthur Andersen's Lessons from the ABM Battlefield* (New York: Mastermedia Ltd., 1995).
7. Andrew Grove, *High Output Management* (New York: Random House, 1983), p. 173.
8. J. LeBlanc and Tom Roberts, "ABC Assists in Building a More Profitable CASE," *As Easy as ABC: ABC Technologies Newsletter* (Fall 1997).
9. S. Schaefer, "ABC Implementation: Planning Is Everything," *As Easy as ABC: ABC Technologies Newsletter* (Spring 1991).
10. Peter M. Senge, *The Fifth Discipline* (New York: Doubleday Currency, 1990).
11. Steven Covey, *The 7 Habits of Highly Effective People* (New York: Simon & Schuster, 1990).
12. Thomas D. Davenport, "Information Behavior: Why We Build Systems That Users Won't Use," *ComputerWorld,* September 15, 1997, p. 3.
13. Faheem Zuberi, "Sallie Mae Puts ABC to the Test," *As Easy As ABC: ABC Technologies Newsletter* (Spring 1998).
14. Rapid Prototyping is a technique formalized by ABC Technologies Inc. to accelerate ABC/M learning using real-life examples and on-the-spot cases. Participants spend two to three days working together to build a model of the enterprise on the fly while learning how to understand and use an ABC system. Cross-functional management teams usually embrace this concept as they can learn together and gain a quick understanding of model building and decision rehearsals. Usually ABC consultants work in conjunction with a software specialist where the consultant organizes thought in the room while the keyboardist organizes the model design and input.
15. Eileen Morrissey and Gary Hodson, "A Smarter Way to Run a Business," *Journal of Accountancy* (January 1997), p. 48.

9

SECOND WAY: UNDERSTAND THE EVOLUTION OF ACTIVITY-BASED INFORMATION SYSTEMS

Previous chapters discussed the differences between ERP (enterprise resource planning) systems and ABIS (activity-based information systems). This chapter introduces activity-based players who can help in efforts at ABC/M. People entering this market probably are not going to implement an ERP system to do ABC/M. They may be in the midst of an implementation that requires them to ask the basic questions:

- Should we purchase a stand-alone system or go with an ERP?
- How do we know a good ABIS from a bad one?
- Whom do we look to as we find our way into this market?

All these questions are valid and are addressed in the following chapters. First it is important to know how events or systems came about. Armed with this knowledge, the practitioner will have an effective basis for making the decisions toward a great implementation.

IT IS A JUNGLE OUT THERE

Since the 1980s, activity-based systems have challenged the status quo of the legacy cost management systems. The battle has been fought and won on concepts, but now a new generation of information systems is dotting the landscape. In some ways, this new generation of information systems owes its growth to history.

IN THE BEGINNING, GAZELLES ROAMED THE LANDSCAPE

Fueled with new changes in the way the manufacturing organizations viewed overhead, several academic thought leaders developed their own software packages. These systems were designed to prove a point rather than become integrated production-quality systems. Like gazelles, these academics roamed the landscape, leaping higher than others in attempts to pull their audience to new heights of knowledge. Without these players, there would have been no market.

GIRAFFES INCREASED THE VIEW

Very soon, Big 5 consulting firms acquired, adopted, or built their own packages to meet the clients' demand for implementation systems. Like giraffes, these consulting firms and other key individual consultants predicted the demand for ABC and ABIS. They could see the landscape. These systems were first embedded within large and small consulting firms, and the market fragmented with methodology-specific systems. Hence companies interested in implementing ABC/M would purchase thought leadership and software from the same firm. Note that several smaller firms had much to do with the growth in the marketplace.

From this scenario has evolved the multistage, multilevel information systems marketplace. These early participants fueled and created distinct waves in the ABIS marketplace:

- The founding parents of ABC/M proved the concept by continuously implementing systems and proving their point.
- The knowledge brokers, namely the consultants, built and borrowed technology to prove that systems can be implemented effectively.

THEN CAME THE SOFTWARE TIGERS

A new breed of professional software vendors formed in the 1980s, vendors who were not finance literate or thought leaders but software developers who had identified this market as a growth opportunity. They used PC technology, price/performance, and mass production strategies to gain mass appeal. Their credo was more oriented to the off-the-shelf, easy-to-use software. A few such vendors are ABC Technologies Inc., Armstrong Laing, and Sapling.

These and other independent software vendors pushed the envelope, introducing a whole new focus to the marketplace—one that said others could achieve the same success. During these high-growth market years, the audience was beginning to evaluate proof-of-concept and pilot stages. Few true production systems existed.

Several Big 5 firms began to move away from offering their own software packages. Over the years, many of these firms decided to focus their practice on pure consulting. They began to work with any package that their customers wanted. A few, like PricewaterhouseCoopers LLP still developed and offered their own software to the marketplace.[1]

STRONG AND BOLD BUFFALO DECIDE TO JOIN THE PACK

From 1997, the larger integrated emterprise and database system vendors, which include SAP, PeopleSoft, and Oracle, declared that they would offer solutions in the ABC arena. Like the bold buffalo, they are strong, focused, and credible.

Today, what used to be viewed as separate and distinct markets are converging:

- Data warehousing
- Activity-based analysis
- Analytical applications like budgeting software, consolidation software
- Desktop navigation tools
- On-line analytical processing applications and servers

These technologies are all participating in information gathering, analysis, and deployment. As usual, all these changes are fueled by technological advancements and strong need to find meaning in data in corporations.

SURVIVAL OF THE SPECIES

Thought leaders—the educators who discovered and defined this space—who drive the industry are like gazelles who leap high and move faster than anyone else. As thought leaders, they are also faster than others in predicting the future.

Consultants, like giraffes, love to deal high in organizations and can see better than most. They have helped the market expansion by getting to the decision makers and influencing their ways of thinking. However, they are nomadic, constantly searching for interesting challenges. They must not be assumed to be full-time employees. In fact, their job is to cause change for the good and to transfer knowledge.

Software vendors, like tigers, must be handled only by those with training. These nimble cats are swift and respectful of the buffalo, the ERP providers, who are an animal that takes charge and is focused. They seem less nimble than the tigers but bring legitimacy to the marketplace.

These analogies help in identifying the personality of the generic player in the industry. No business decisions should be made based only on these descriptions.

WHAT DOES THE FUTURE HOLD?

Understanding the past will help in gauging the future. To recap, the industry was born in the minds of thought leaders. From there the evolution seems as follows:

- Academic discovery and communication of the information.
- Consultants use it as a practice.
- Consultants and thought leaders create appropriate software.
- Independent software firms create off-the-shelf software.
- Larger systems providers enter the market.

None of these evolutionary steps invalidates another. They all live in concert with each other and have found their own space and use. However, interesting combinations and partnerships are evolving. New alliances are predicted that will rekindle the ABC movement. All members of this ABM ecosystem seem to want to grow the industry and have discovered that not one of them can do so without the other. For example, in 1998 PeopleSoft announced a partnership with KPMG Peat Marwick, Robert Kaplan, and Robin Cooper to build an ABC solution. In the same year, ABC Technologies signed SAP into an equity partnership and a joint development and marketing agreement, moving the industry to appreciate the balance between integrated ABC/M and analytic ABC/M and the need for both.

DECIDING WHICH SOFTWARE VENDOR TO WORK WITH

Every data sheet tells the same story but is equally exaggerated. Think about a trade show. After visitors are on the floor for a while, all the products and all the sales people begin to look the same. All the vendors seem to say the same thing. Every firm seems to be the "market leader," "dominant," "high performance," and "undisputed choice." The key questions in making the purchase decision are:

- How do we truly differentiate one from another?
- If our needs evolve, how do we buy for it now?
- How do we ensure that our courtship with a software vendor will end in a marriage?

- What do we do if our consultant recommends a vendor against our own instincts?

Clearly, no two software vendors have the same personality. They are different, from the way they build software to the way they market them.

RULES OF ENGAGEMENT IN UNDERSTANDING A VENDOR

Ten rules of engagement can assist in comparing and understanding vendors in the analytic ABC/M environment. They are:

1. Understand the firm's needs.
2. Make a rational decision using a process with checklists and reviews.
3. Evaluate more than software.
4. Do not use the vendors exclusively to learn ABC.
5. Visit vendors and tour.
6. Do not get caught in trends.
7. Always keep in touch with the market.
8. Remember the 10 questions. Consider 10 quick questions to profile a vendor.
9. The nature of the dance is the nature of the relationship.
10. Control the demonstration.

Understand the Firm's Needs

If someone does not know beauty, then everyone looks beautiful. Understanding what the firm's needs are and also its ability to absorb the growing needs of the organization are the keys to any process of evaluating and working with others. Many ABC/M projects really started on the wrong foot—they never knew what they really wanted and so never got it.

Make a Rational Decision Using a Process with Checklists/Reviews

The team evaluating vendors must understand the priorities of these needs and the appropriate trade-off the management team are willing to discuss. Compromise is the essence of the evaluation process. "Process" is not used lightly in this discussion. A haphazard, gut-feel approach will leave room for errors in judgment. Without a practical process to guide evaluation, emotion, bureaucracy, politics, and irrational conjecture will flood the decision-making system.

Evaluate More Than Software

There are few barriers to entry in software. All it takes is an office, software engineers, and a phone to bring a software company to life. Great software companies are more than this. They are systems built to carry the software users through the hard times and to support them through their growing needs for results. In the evaluation, consider that what is being evaluated is not merely a product or a service but membership into a new club. Software is changing all the time; users of any software package will be having a relationship with its maker. Recognize that software is only one aspect of the solution organizations. Other components brought together make the product whole, that is, the service component, the support infrastructure, the company itself, the relationship your company can count on, the financial infrastructure backing the company etc.

The checklist in Appendix A (feature table) provides some grounding with respect to activity-based software information systems.

Do Not Use Vendors Exclusively to Learn ABC

In many demo situations, clients ask basic questions about ABC and how it can be applied. Software vendors will always educate potential clients about the industry and the market but for one reason—so people will buy their solution.

In their education process, software vendors aim to show their product in the best light and to downgrade competitors. Undereducated clients can be led astray.

Teams meeting with software vendors must know what their firms need and what they expect from the meeting. Software vendors are useful learning resources but should not be the sole source of education. The organization should develop a model for educating itself prior to the demo. Knowing this, the software vendor that can educate and be an unbiased source of knowledge should be respected.

Visit Vendors and Tour

Software and systems vendors who are proud of their growth and want to take advantage of their reputation will share everything about themselves. They tend to invite clients to visit their offices and tour their facilities. Many ABC software vendors are small and tend to work out of small offices. Others are really sales offices with research and development overseas. It is important to meet the customer support professionals who will help when the going gets tough. Furthermore, it is necessary to understand their quality assurance procedures, if any. This could save a firm from disaster.

Before visiting vendors, make sure to set the agenda and meet the senior management. Present the firm's strategy and make them accountable to your needs. During the tour, carefully observe everything: Is the sales organization busy? Can engineers answer ABC questions? How many are working on the product the firm is purchasing? Where is the firm's sales representative in the hierarchy of sales?

All in all, the best way to pick a vendor is to visit one and observe their vendor-client relationships.

Do Not Get Caught in Trends

Many ABC/M projects are visionary projects—they are change management oriented. Visionaries lead these projects, and they are filled with the energy of possibility. This is a needed characteristic for dramatic change. However, the very characteristic that change agents possess can be the source of their downfall if not held in check.

In the early phases of a project, organizations are enthusiastic and believe they have the unlimited capacity for technology adoption. Trends in new advances can only accentuate addiction to technology. Web-based technologies and artificial intelligence systems for budgeting are new noteworthy trends. These are valuable technologies of the future, but a firm cannot build for tomorrow on tomorrow's technology. Only today's technology can be used to build tomorrow.

Always Keep in Touch with the Market

Loyalty is a virtue. In ABC/M projects, risk is inherently high and sometimes loyalty to one vendor can be a mistake if the project team does not keep in touch with what's offered in the marketplace. Almost 50 percent of implementations fail to meet expectations. One way to prevent any setbacks in the technology elements of the project is to hedge one's bets by creating alternatives to the technical implementation. For example, say a firm decides to build a model with multiple dimensions. In the course of creating the model, it is revealed that it takes 50 hours to process onto an OLAP cube for analysis. Turnaround time might destroy the schedule. However, if the project was designed so that other OLAP engines could be programmed from the model, another one might take less time. To ensure portability of the modeling environment, more than one vendor must be available in case trouble brews.

Keeping multiple seconds and one loyal partner is difficult, especially since everyone is trained on a specific technology. Keeping in touch with the market can only reduce the risk, not eliminate it.

Remember the 10 Questions

The following 10 questions will help profile an ABC/M vendor. Consider that if the vendor is large with many tool offerings or smaller, focus and depth in ABC is essential. Sometimes, testing the larger vendors with diverse tools on focus is useful while testing the smaller vendors on the ability to scale upwards is equally important.

1. Is the ABC vendor more than a one-product company or a one-product group with no diversity in their ABC offering?
2. Is the ABC vendor a software company or a consulting firm in disguise?
3. How large is the vendor's installed base in ABC/M?
4. Does the vendor support the ABC products with training and technical services?
5. Does the vendor have support of an active, international user group? Does the vendor have regional ABC/M user group meetings?
6. Does the ABC product adhere to industry standards?
7. Does the ABC product support an open interface?
8. Does the product have a migration path to higher performance systems?
9. Does the product include or interface to other productivity tools like process modeling, or performance scorecards?
10. Does the vendor support an integrated ABC environment?

The Nature of the Dance Is the Nature of the Relationship

The customer-vendor courtship is much like a marriage courtship. The first date is a wonderful experience for many, but can it last? Through this courtship, firms will come to understand the style of the prospective partner and how it views a partnership.

Some vendors like selling futures while others only sell the present and never talk about the future. Others are hungry enough to work with firms regularly while others will disappear when the purchase order is cut. Some keep commitments while others promise but deliver late. Does the firm want a vendor who accepts its idiosyncrasies or one that challenges them?

Throughout the process of meeting, understanding, and purchasing software for ABC, an organization must assess the value and the future of the partnership. As we all know, partnerships are always deemed to be strategic even when they are merely tactical and short term. The selection of a vendor for software must be categorized one or the other and treated that way.

Control the Demonstration

Many demos are considered the ultimate decision point. Realistically, they are like beauty contests when they should be like fitness tests. They are staged events that make or break a sales opportunity. This emphasis may be due to the seeing-is-believing mentality in the buying process. Yet what people see may not be what they get. Often demos are controlled by the software providers. They follow tight and fail-safe scripted demos that highlight the power of their solution and loosen a firm's mental grip on their competitors.

The ultimate lesson to learn is to control the sources of information used in a demo, control the issues discussed, control the decisions to be made, and control the attendees' expectations.

The following checklist indicates how to organize the meeting based on requirements and expectations.

- The number of demos and vendors to review
- Understand their strengths and weaknesses ahead of time and communicate
- The areas to be highlighted—"the show me's."
- Who should attend and what areas are they to observe?
- How flexible are the demo-givers to changing the issues they set?
- Measure their consistency with their literature and Web-page information
- Questions to be asked. The most important question to ask is about the demo-giver—ask how long he or she has been with the company. The answer may be surprising.
- Provide a base model for the demo-givers to build and inject an error to observe their response.
- Provide each of the demo-givers scores. Brave clients may share them with the demo-givers.

- Don't react; anticipate their actions and put them into the agenda for testing.
- Respect the vendor. Above all, don't make the demo a selling event. Make it a dialog to understand if two mature parties can see a common future.

NOTE

1. Historically PricewaterhouseCoopers LLP maintained their own software package. As of May 1998 Oracle Corporation picked up this package as an offering in their suite.

10

THIRD WAY: KNOWING THE ROADMAP FOR DESIGNING AN ACTIVITY-BASED INFORMATION SYSTEM

Businesses have found that ABC uncovers tremendous opportunity for optimization of costs and invariably for directing strategic decisions for which cost is a component.

Previous chapters focused on understanding what is necessary for successful implementations. Without understanding the true potentiality and strategic position an activity-based information system provides, many programs have fallen short of expectations.

Although the ABC/M industry is still populated with proof-of-concept pilot programs, several pioneering organizations have ventured into the world of "continuous and sustainable" ABM. In observing successful programs, certain key phases of an ABC project emerge.

One major factor that enables continuity is deploying technology which is manageable, scaleable, and repeatable. Another factor involves the process and people components—that is, a strong project orientation to the program. CASE Corporation of Wisconsin began its implementation in 1991.[1] In 1993 it reaffirmed that its approach provided results. A key learning for this "case" is that, early on, it was decided that a phased approach to implementation was best. A phased approach is invaluable due to the dynamics of many ABC endeavors. First, many ABC teams are populated with multifunction members who have other primary functions. Some ABC projects can last years; a phased approach, with measured deliverables, creates signposts for success as well as allows team members to know when to move to the next phase, sometimes with different people. It also permits them to spend less time on the rules of engagement and more time generating results as all the rules that run the project are defined.

POLITICS OF AN ABC/M PROGRAM

There are as many ways to manage an ABC/M program as there are personalities in one. While most programs hunt for the ultimate "aha" with data gathered over a pilot program, those that succeed have other common traits. One strong trait of successful ABC/M programs is project management and scheduling. A carefully planned project has a greater chance for success and also a greater chance to be flexible and to bring along the team that takes it on.

Here project management does not include all the strategic elements that go into the creation of an ABC/M project. Good technology project management can transform objectives into software models and then into reports that can assist users to make better management decisions. To ensure that, the end result of a technology project management process is the creation of a program that is scaleable, manageable, and repeatable. To define:

- "Scaleable" refers to the ability to grow the architecture of the endeavor and to bring larger and larger amounts of data, build ever growing models of the enterprise, and report to an increased amount of people.
- "Manageable" refers to the ability to contain both the physical work and the cost of keeping the system going.
- "Repeatable" refers to the ability to replicate the ABC/M environment in multiple locations using the same techniques and the same technologies.

PHASES OF AN ABC EXERCISE

Exhibit 10.1 outlines distinct phases to a typical ABC/M exercise. What usually tends to happen is:

- Objectives are given top-down from management and the sponsor of the program, and the project is staffed and begins.
- The team begins by swarming the operational teams with questions and questionnaires.
- Educational meetings begin as these questions face resistance and fear.
- The enthusiasm grows as more and more discover people that the project that will solve the common cold has begun and that reports will be sent to management for change to begin.
- The data-gathering phase begins with many people being interviewed about their jobs and what they do to help others.
- The team grows silent as it moves into the data integration phase, where members and IT professionals collect other data for their models.

PHASES OF AN ABC EXERCISE

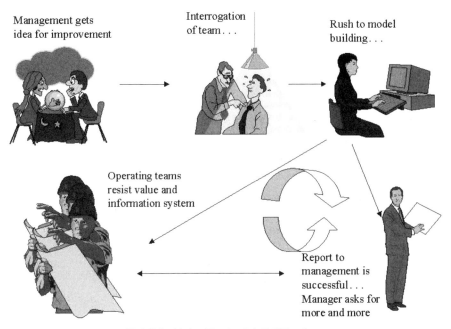

Exhibit 10.1 Typical ABC Project

- Now the model is ready to be built, and management see progress while they are fielding cautious questions from the operating teams who say that they are not sure about the information.
- Months have gone by and the project team is hard at work building the model. Finally they are ready for the presentation. The information is only a month late.
- The presentation to management is a roaring success. The managers give the team a standing ovation and multiple plaques to hang on their cubicle walls.
- As the clapping dies down, the managers want to make adjustments to the model and have more questions arising from the exercise than before. The team dares not tell them that they put this model together fast and some of the questions may require them to redesign the model and gather more data. Furthermore, three members of the team must return to their real jobs.
- Management has seen the opportunities and is now blinded by enthusiasm. They send the team to produce more exciting results.
- The team approaches the same data sources and interviews with the same vigor. Meanwhile the operating teams have heard nothing except that the

meeting was a success. They are weary of further questions and answers. They want new information.
- Once again, the project team presents to management. In horror, the operating teams exclaim that the data they obtained were only estimates and that the results must be wrong. In fact, things have changed in the last six months. They challenge the entire process and the methodology as well. Furthermore, they now realize that they will be using PC-based systems to interrogate this model and obtain answers to the most pressing business questions. They are not trained on such a financial system and realize that they need more resources.
- The IT group hears of this project in a meeting and now declares that they must certify the technology and ensure that no corruption of information in their data stores has occurred. Not only that, they believe that the technology used is inferior to the choice they would have made. To support this, they must get funded.
- Meanwhile, management is waiting for answers to new questions and are beginning to believe that the ABC/M team leader lacks the skills to continue the project because of all the complaints they are hearing. They say, "Maybe this stuff called ABC isn't real."
- The project team now states their challenges:
  —A resistant team
  —A nonscaleable and nonmanageable model
  —An unintegrated application environment
  —A lack of funding and resources
  —A plan to revamp the system and produce results that will double the time for the pilot
- The program is delayed until further notice and the teams continue with no changes.

ABC PROJECT WORM

Too many times, ABC/M endeavors of this type end in the loud roar at the finish line followed by the solemn silence. Exhibit 10.2 illustrates the eight phases of an ABC/M project that can produce continuous effectiveness. They are:

1. Objective setting
2. Data gathering
3. ABC modeling
4. Integrating

5. Reporting
6. Empirical data replenishment
7. Forecasting and budgeting
8. Re-creating reports

Each of these phases have distinct activities and deliverables. Any one of these phases performed incompletely or incorrectly can result in project stagnation or failure. Note also that these phases are shown in serial order for simplicity; some of these phases can be developed concurrently.[2] Inherent in this project guide are the following:

- Projects never end and are continuous.
- Checks and balances exist in each phase to ensure proper handoffs to the next phase.
- Documentation is vital in each phase, and an overall project plan must be developed prior to launch.

Objective-Setting Phase

Hewlett Packard North American Distribution Organization distributes 21 product lines of printer products and achieves $7 billion in product distribution to over 300 resellers.[3] The firm entered the world of ABC searching for:

- What channels are best?
- Which customers are best?

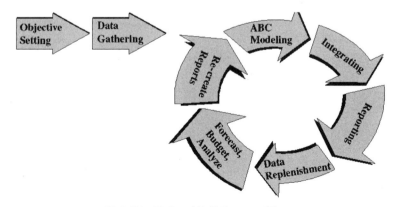

Exhibit 10.2   ABC Process Flow

108 KNOWING THE ROADMAP FOR DESIGNING AN ACTIVITY-BASED SYSTEM

- What types of service are best?
- How do they price these services?
- What is the optimum supplier-to-reseller configuration?

This is a set of objectives that an ABC/M project can achieve, given set resources and time. Note that they are concise, definable, and measurable. The most important phase of any project is the objective-setting phase, because it is here that mistakes are made in understanding the boundaries of the project. It has been said that the "finesse of using ABM is primarily related to identifying and managing scope."[4]

The excitement of a project opportunity sometimes can be so high that the discipline of defining the challenge is overlooked or washed over. But it emerges with a vengeance at the worst point of the cycle—when the reports are due and all the money is spent. New Zealand Post (NZP) created a strong project with predefined objectives that resulted in success.[5]

NZP drove an implementation in 1996. At that time it carried 774 million post letters, 15 million fast post letters, 971 post shops and centers, and 27 mail centers and employed about 9,300 people. The objectives with ABC/M were to:

- Provide generally acceptable costing methodology to assist NZP in understanding cost of post processes and products as well as setting external interconnect and transfer prices based on an understanding of those costs
- Benchmark
- Determine cost trends

New Zealand Post broke the project into two phases and by business groups. The project team was comprised of three external consultants, four finance staff, and three "letters" representatives. A steering committee watched and guided the team's activities. This team consisted of the CFO as head and representatives from each business group. NZP is also a strong user of OLAP technology and multidimensionality.[6]

In 1991 Dayton Extruded Plastics began its ABM implementation using an advisory team, a set of objectives and operational milestones based on classic project management techniques.[7] Team members were selected according to their areas of influence, the levels within their organization, and their time available. Outside consultants were relied on only for team training, periodic advice, and theoretical knowledge transfer and education. Dayton wanted to own the systems and their implementation. Dayton's functional objectives were far more detailed and exhaustive than those for NZP. They included:

- Understandable by information users. They would distribute the systems and the information rather than use a centralized repository.
- Life-cycle costing would be included.
- Managers would own their own designs.
- Activity information would be included in addition to costing products and customers.
- Different levels of aggregation would be achieved by micro- and macroactivities. Hence different levels of managers would view the information.
- Cross-functional reporting would be accomplished using attributes. Here they would tag activities with attributes, such as cost-of-quality attributes.

While certain organizations develop objectives for the project, others recognize that their current systems are producing the wrong information. Oregon Freeze Dry realized that its conventional cost system was "stale." In its own words, "it wasn't able to reflect the growth and product diversity."[8] In this case, objective setting is clear—get to the right information.

Data-Gathering Phase

The first time a project is launched, everyone is curious and forgiving when data are being gathered. It's the second, third, and fourth time that the problems begin when the participants get tired of the questions and of the data-gathering work. But there are inherent challenges even the first time. As Harris Corporation understood, "for the most part, there was never one expert on all the data available."[9] The type and nature of data being gathered are diverse and can reside in various places in the organization. (See Exhibit 10.3.) In some ways, the very value of integrated information is the challenge in an ABC/M project. There are three types of data being gathered:

Types of Info	MARKETING	SALES	ADMIN	MANUFACTURING	MANAGEMENT	ENGINEERING	CUSTOMERS
Driver data							
Quantity							
% Time spent							
Margin info							
Profit info							
Activity info							
Resource info							
Sales table							
Costs							
Etc.							

Exhibit 10.3 Who Is Responsible for Information?

1. General ledger information
2. Empirical information—information found in individuals' heads, such as percent of time spent on a particular activity
3. Operational information—related to driver information—number of customer calls or manufacturing information about shipments

Specific to developing a model, data take the following forms:

- Assignment information—information about source and destination relationships
- Driver data—types and quantities of activities and resource drivers.
- Resource/activity/cost object data
- Attribute data—any tags placed on resources, activities, products, channels, bill-of-materials information, sales information

There are several ways to gather data:

- Interviews
- On-line surveys/data-gathering systems
- Observations
- Classroom ad hoc question and answer periods
- Focus groups
- Paper surveys
- Time cards

These methods have their advantages and disadvantages. When considering any of these methods, consider the amount of time required and the level of detail truly required. In many ABC/M projects, this phase is overperformed and too much data are collected. New Zealand Post used on-line data capture, visiting 27 mail centers to install data capture systems, then collecting and analyzing dynamic data and snapshots. These results were entered into an Access database and Excel. While creating 7 resource drivers and 14 activity drivers, they collected this information from 11 different systems. They were challenged with not only the time-consuming data collection but also faced the different terminology and mapping problems between systems. Data gathering is not system integration, which is discussed in the next section.

US Airways realized that indiscriminate data gathering would lead to disaster. So it was noted that "one of the first steps this cost management team took was to determine a series of primary questions and issues, the answer to most of which drove model design and detail."[10] This team focused their data gathering by identifying the four key business issues that drove the activity collection cycle.

Citymax learned that when it started to talk to people about activities it performed, interviewees were surprised. They had never been asked what they do on a day-to-day basis.[11] All in all, the data-gathering phase of the project must be well planned and executed. If it is assumed to be a one-time event, it will be, and the project will never cycle to improvement. In planning, answers to the following questions should be composed[12]:

- How can we make it easy for data to be available? (Automation can play a strong role in this area.)
- How can we improve the quality and accuracy of the data? (Inconsistencies of data across systems are the greatest impediment to speed.)
- To what extent is history required?
- How important is data replenishment? (It is critical.)

Integration

Almost 70 percent of information for an ABC model is not input through the user interface. Hand-keying information is not necessary, as information needed resides somewhere on some system in the organization. There are two ways to import information into the model (see Exhibit 10.4):

1. ACSII importing of information
2. Direct importing of data elements using a query engine that grabs information in specified formats from many systems

Importing information from heterogeneous systems is seldom the challenge. The challenge seems to be in cleaning up the raw data into a consistent form. Sometimes it comes merely down to ensuring that "ABC" not be input as "ABC" or "A-B-C" or "A-b-C" or "BCA." Although this is a simplistic example, inconsistencies like these can destroy the project schedule.

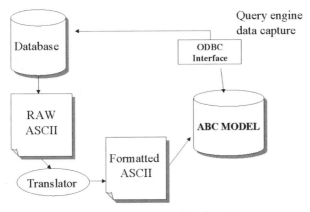

Exhibit 10.4 Importing Strategy

This phase is very important if your project is to last past the pilot stage. With integration your model can be replenished with direct connections to data sources. This does not take away the need to gather information found only in people's heads, that is, empirical information: "70% of the labor that goes into maintaining data repositories is spent reconciling and cleaning up data," states Alan Paller, director of education and research at the Datawarehouse Institute in Bethesda.[13] In a nutshell, integration is not merely connecting the data sources to the data repositories in the model. An intermediary phase (see Exhibit 10.5) of data cleansing and data staging exists.

Modeling Phase

Often the model phase is perceived to be the most important one, but the experienced ABMer may challenge that perception. As mentioned before, the most important part of the ABC/M exercise is in its design.

As in any design project, 80 percent of the costs fall in the design phase. Armed with this undisputed knowledge, many ABM projects spend the least time and energy in design. Consequently, they spend more time undesigning their design. Steven Covey asks us to think with the end in mind. This mantra is the mantra of the experienced modeler.[14] For example, "Let's build models only after we architect them based on the questions we want answered" would be the obvious path to model nirvana.

Fortunately and unfortunately, the most success that can be gathered from an ABC/M pilot is to attempt another endeavor to answer the next set of questions that the pilot has generated. That's the true power of ABM—the right questions. The designer must model and architect a nonstatic model—one that is iterative and yet scaleable. Hence there is no perfect way to model—just steps to the ultimate architecture.

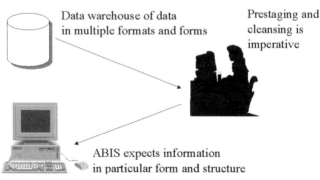

Exhibit 10.5   Data Staging and Cleansing

Model Building Tricks of the Trade

When it comes down to the project, the more knowledge users have about the task at hand, the better their chance for success. The following ideas and hints are designed to assist in model building and design.

- *Don't be guided by how the software works. Don't jump into the software and begin modeling.* ABC Technologies decided to eat our own dog food and carry on an ABC project in 1995. The team of executives seemed so busy with the day-to-day that they could not start. The CFO was excited by the idea of doing ABC but did not have the resources to start. A consultant was hired to get the troops on the target. At the start of the project, the team built a repeatable list of activities—an activity dictionary, which outlined all the legitimate activities, their interconnects, the associated drivers, and performance measures. Several organizations build their models this way. (See Exhibit 10.6.) Such a dictionary provides attribute definitions and process membership descriptions for core activities in an organization. From this point, model design becomes less complicated and fully documented.

Name	Deliver training
Process	Order delivery process
Definition	Deliver public and on-site training for paying customers
Inputs	Schedule training classes
Cost Driver(s)	# of training courses # of training course sessions # of students trained Type of class Type of venue
Performance Measures/ Outputs	Student Evaluations # of training course sessions # of students trained
Driver(s)	# of students trained
Characteristics 1. Performance leverage 2. Customer value 3. Quality deployment 4. Primary/secondary	benchmark differentiates prevention primary

Exhibit 10.6 Activity Dictionary Page Sample

- *Build models for a purpose.* Teams must decide and communicate the limits of their model building and ensure that the clients of this information are aware of what they will receive in the first step. Teams can achieve this by understanding and communicating the answers to the following questions prior to modeling:
  — Is the model designed for strategic or operational decision making?
  — How deep and wide is the model going to be?
  — Does the model go to the product and service levels, or is just activity analysis being done?
  — What can be done to this model that is being built after it is complete?
  — What information is needed prior to building this model and in what form?
  — Can this model be replicated in other locations in the company and will the activity dictionary work?
  — How much education is needed to put the model in place, get it off the ground, and keep it up?
  — Have integration issues been faced early so that model inputs will not have to be re-created?
  — What trade-off can be made with respect to the competing interests between accuracy and time taken to model?
  — If a prototype is being built, is it repeatable?
  — Have the right design methods been selected?

  Once model building starts, the requirements and expectations begin to expand as excitement grows. So, project management becomes even more crucial. Chapter 11, which captures all the basic elements of controlling a project, discusses more guidelines on project expectations.

- *Link models to strategy.* Most ABC/M projects may complete pilot programs but fail to last because of the lack of upper management support. The reason upper management cannot see their way among the trees in the forest is because they are not familiar with the model. If the ABC/M exercise can be related to strategic issues that management is struggling with, the leaders will have a newfound interest in and bring a new drive to the project. A leading officer in an international government agency once declared that if an activity did not align with a strategic theme, he would either remove this activity or decide that it is not at the highest level and must be embedded into another key contributing activity. This way he aligns strategy to activities and keeps his model(s) at the right level of complexity.

- *Milestones, measurements, and management.* Within the model-building exercise, ensure that there are miniproject milestones and measureables. Exhibit 10.7 illustrates one such detail in building a production model. Those who measure by the mile can slip only by a mile.

ABC PROJECT WORM

Task	Responsibility	Target Date	Actual Date
Kick-off project team			
Rapid prototyping class			
Draft project plan			
Complete model 0			
create sales table			
create driver dictionary			
create activity dictionary			
develop communication plan			
Define and generate reports			
Update plc documents			
ABC team review of update			
Management review			
Incorporate feedback from mgt			
Resource module design review			
Import general ledger information			
Assign resources			
Activity module design review			
divide subprocesses into activities			
revise and update drivers			
collect and input driver quantities			
assign activities			
update activity and driver dictionary			
Cost object design review			
determine level of detail			
revise and update drivers			
collect and input driver quantities			
finalize sales table			
assign cost objects			
Manager review			
Attribute design and implementation			
Develop reports			
Present results to management team			
Ongoing modeling and reporting			

Exhibit 10.7 Sample Preliminary Schedule

- *Build smart models, not big models.* Companies are moving rapidly from pilot proof-of-concept installations to enterprise-wide production systems. It is encouraging to note that the ABC/M industry has moved beyond pilots that are discrete events to more on-line, active analysis and reporting systems, which are more continuous. This evolution has its pitfalls. One way to avoid these pitfalls is to build models correct-by-construction, that is, there are ways to build models to avoid problems with not being able to increase its functionality or complexity. In the postpilot phase of model design in the early 1990s, models crept up to about 2 to 5 megabytes in size. Today it is customary to be dealing with models that are 90 megabytes to 1 gigabyte in size. Models' sizes are increasing exponentially, and software providers are working hard to ensure that size is never a barrier to design. Is this a good thing? Model complexity is doubling every two years, and with it comes the assumption that effectiveness is doubling as complexity doubles. In fact, as complexity mutates, organizations may witness a gradual degradation of effective decision making. The true measure of an ABC/M project is not how much is modeled and not even the decisions made to improve the enterprise—it is the mean time between decisions (MTBD). If the model cannot be turned around to address decisions fast and flexibly, the organization loses in this market.

This complexity is driven by the increasing demands placed on models. As the models are asked to do more, they begin to creep into incredibly accurate and complete creations that lack flexibility, maintainability, and scalability. Four areas of optimization should be considered prior to actual modeling:

1. Resource modeling
2. Multimodel strategies
3. Activity modeling
4. Driver selection

1. Resource Modeling

Many modelers realize that detail accounts from the general ledger can complicate the model from the start. However, most modelers want to have the detail captured and traced back to actual general ledger accounts. Hence, these competing interests seems to result in associating one resource element to one general ledger element. (See Exhibit 10.8.) This introduces tremendous redundancy just in drivers and assignments, because many accounts have the same driver. Any attempt to reduce complexity, as shown in Exhibit 10.9, by consol-

ABC PROJECT WORM

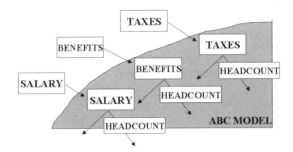

- One resource account per detail account
- Detail captured in model
- Redundant drivers and assignments
- Detail reporting becomes difficult

Exhibit 10.8 Resource/Detail Mapping

idating resource elements prior to inputting information into the model would reduce the ability to trace information back to general ledger accounts. It may not work as it compromises the model.

Detail must be captured in the model only if it is believed it must be reported at this level. If it is needed and general ledger reconciliations are expected, consider the use of resource hierarchy in the software. Hierarchy is a way to preserve the integrity of the model but not increase its complexity. (See Exhibit 10.10.) Also, hierarchy optimizes drivers and assignments.

2. Multimodel Strategies

Several organizations want to incorporate ABC models into their information infrastructure. They think of these models as archiving media rather than deci-

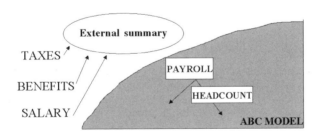

- External summary of detail accounts
- Detail lost prior to model
- General ledger reconciliation complicates and adds step
- Detail reporting lost

Exhibit 10.9 External Summary

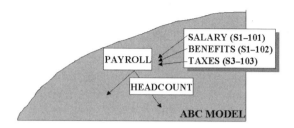

- Introduce resource hierarchy-cost elements
- Detail captured in model
- Optimized drivers and assignments
- Optimized detail reporting

Exhibit 10.10 Resource Hierarchy for Detail Mapping

sion support tools. Most organizations have an abundance of irrelevant information, and the tendency is to put all this information into a knowledge engine, such as an ABC/M model, so that the model will decide the value of the information. (See Exhibit 10.11.) Consequently, models become large and unmanageable.

Lean models are a symptom of a mind-set—don't overcomplicate, separate! Classical computer science reminds us to break larger parts into subparts and decompose a challenging task into minute tasks—they call it "structured decomposition." A general ledger is an archival system that stores all possible information pieces, dated or otherwise. When people assume that the ABC/M modeling environment is not an archival system but a period-based analytical

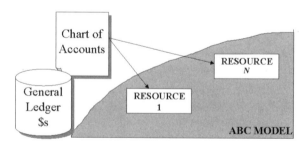

- Chart of accounts drives resource detail
- Period detail captured in single model
- Unused accounts default to zero
- Single model for YTD/MTD summaries
- Inefficiencies propagate throughout model

Exhibit 10.11 Single-Model Strategy

tool, they will look to the general ledger for current cost and resource information rather than grabbing everything available in the hope it can be used. (See Exhibit 10.12.) Another way to view this periodic information is to develop several models and develop consolidation methods for reporting.

Managing these models may be challenging, but the simplicity of each model can be maintained. Because the structure of the model may change periodically, maintaining a single model will require redundancy in the model and the acceptance, at any period, to hold inactive accounts and structures.

3. Activity Modeling

The first question that comes to mind when dealing with activities is "how many?" The next question seems to be "At what level?" As was mentioned earlier, the more important result of an ABC/M project is understanding not costs but the *impact* of activities on costs. Since ABC is targeted to operational managers, who value detail, activity detail will be demanded. However, too much detail and too many activities will kill the exercise. (See Exhibit 10.13.) How does a modeler maintain the integrity of the exercise and still satisfy the insatiable desire for detail?

Exhibit 10.14 shows a hierarchical modeling methodology. Within the same module—the activity module—macro- and microactivities hold information. With this process-task relationship, redundant drivers and assignments are reduced. Note that if all these activities were separated, then drivers would be created for each of them directed toward cost objects, compounding the amount of detail. Operational managers who desire detail can be satisfied while the macromanagement level is adhered to when needed. Use of hierarchy is a powerful method to reduce complexity.

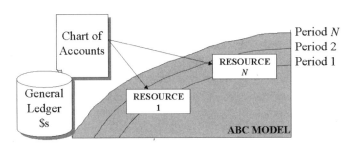

- General ledger activity drives resource detail
- Separate model for each period
- Multimodel reporting for YTD/MTD
- Each model optimized for period data

Exhibit 10.12 Lean Model Strategy

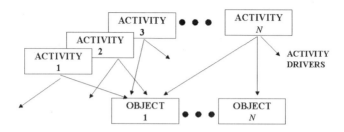

- Detailed activities are key to operational effectiveness
- Activity detail compounds complexity
- Too many tasks/micro activities confuse management

Exhibit 10.13 How Many Activities?

4. Driver Selection

When modeling the enterprise, there is no greater driver to complexity than drivers themselves. Many times driver data are imported from various data sources in an enterprise.

Multiple drivers may seem unique, but if they are correlated—if, for example, they trend in the same direction—they provide very little significance to the model(s). Exhibit 10.15 shows a set of activities, probably derived from different functional areas. One activity uses number of sales orders; another uses number of purchase orders; and the third uses number of invoices. All three of these drivers are almost the same and correlate in movement. Of course, these separately identified drivers will help operational managers see themselves reflected in the model. Even though these drivers are correlated, these driver quantities have to be collected every period.

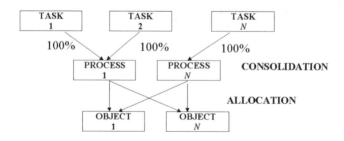

- Model should capture activity detail to task level
- Model should capture few key processes
- Process/task network yields significantly fewer assignments
- Process/task network simplifies analysis

Exhibit 10.14 Macro/Micro Activities

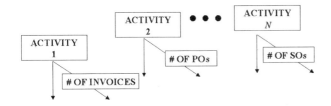

- Multiple drivers compound model complexity
- Product costing minimally affected by driver selection given correlation
- Multiple drivers help influence behavior
- Data collection is most painful part of ABC/M

Exhibit 10.15   Import Complexity-Driver Selection

The integrity of the model and the need for detail can be achieved as shown in Exhibit 10.16 when using one driver for all the correlated drivers. Attribute flags or performance measure tags can associate these drivers together so that all the necessary reports reflect the familiar language. Only a small, finite number of drivers actually perform in a noncorrelated fashion.[15]

Take, for example, a model as shown in Exhibit 10.17. Here, three assignment paths have been created reflecting actual values of assignments from an activity module to cost objects 1, 2, and 3. If we now know that the number of X spent on activity 1 is 10,000, 0, 0.0007 why does the model have to respect the smaller, comparatively insignificant assignment paths that hold 0 to 0.0007 values? Yet many models do so for fear that someday, somewhere, sometime, this would be a significant issue. Yet it may still amount to millions of dollars we may claim! Still, the comparative significance of this value with respect to the others renders its assignment insignificant.

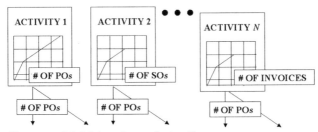

- Use uncorrelated drivers for product costing
- Use performance measures for behavior modification
- Hypothesis: only 7-15 truly uncorrelated drivers
- Fewer drivers yield faster imports and smaller models

Exhibit 10.16   Balancing Objectives Drivers vs. Performance Measures

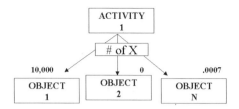

- Assign-to-all significantly increases complexity
- Zero driver quantities add no value
- Scrutinize drivers with high standard deviation
- Electronic links do not differentiate quantity vs. quality

Exhibit 10.17   Driver Data Variation

A number of models that import driver quantities automatically become uncomfortably complex and large. Optimization methods and tools are available to isolate wide deviations in value among assignments of value. For example when a assignment has a quantity of .00007 while another is 10,000, significance is obvious. Removing the smaller assignment from the model for calculation purposes may speed up calculation times significantly. Electronic links are very powerful tools to reduce the modeling time, but indiscriminate importing of data with no regard to their quality and significance is dangerous as they may perpetuate zero assignments like these and build huge models that may be difficult to calculate.

Electronic Data Replenishment

Collecting information from individuals is easy the first time. They are excited about the mystery of ABC/M. They may spend hours answering questions about the work they perform. Yet their patience runs dry the second and third times. If the project is continuous and information on data and the structure of activities changes, data gathering is a challenge. Note that the more important and challenging information is not what is obtained from other systems. Much of the model information is found "somewhere" on some system in the company. A surprisingly large amount of unrelated information is collected. Empirical information is the challenge—information found only in people's heads. Electronic data replenishment is composed of both information found in heads as well as those that reside in systems. Both must be replenished regularly and both require preplanning in an ABC/M project. Data collection systems—electronic surveys or query engines—permit this. Some tools must be able to survey many people and consolidate the information for viewing. Others gather model, driver, and activity data as well; still others include changes in structure and can update the model interactively.

Reporting

Reporting is sometimes the least noted and the most important part of the exercise. ABC/M projects have failed because models were built with little testing as to whether certain reports could be developed from them. There are several ways to report findings to management:

- Use the built-in reports in modeling engines
- Use OLAP navigation engines to build software reports that can be drilled down into
- Use report mining systems that trigger reports on certain predetermined high- and low-water marks of performance[16]
- Use custom reports generated by staff
- User Web-based HTML reports that can be posted regularly for review

Forecasting and Budgeting

Thus far all the work performed has served to build model(s) of the enterprise but users are still running forward looking back. Any results presented will be of the past, probably of the last quarter. Decisions can be made for the next year or period using this information, but there has been no prediction of results and trends of what would happen if certain variables were changed. That is fundamentally what forecasting can provide. Predictive tools are evolving in ABC/M. Several new advances are on the horizon and will serve the ABIS market well. To mention a few:

- Activity-based budgeting systems allow ABIS to run the model in reverse from activities to resource demands. (See Exhibit 10.18.)
- Target costing systems allow ABIS to run the model in reverse from cost objects to activities to resource demands. (See Exhibit 10.18.)
- What-if systems allow variables in the model to be adjusted to view outcomes.
- Process simulation and modeling a time-based and an activity-based simulation of the enterprise in search of impediments to the process, such as bottlenecks.

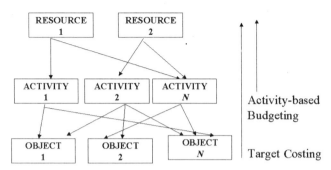

Exhibit 10.18 ABB/Target Costing

PROJECT MANAGING THE PROJECT PHASES

One of most common observations in ABC/M projects is that project management can save days and produce results. Anyone who champions an ABC/M project assumes that he or she will project manage. Each phase in the program's life cycle lends itself to project management. Traditionally, almost all ABC/M projects have some form of control, such as schedules or management meetings. Successful programs utilize significant and noticeable characteristics in project management, which are discussed in Chapter 11.

NOTES

1. J. LeBlanc and Tom Roberts, "ABC Assists in Building a More Profitable CASE," *As Easy as ABC: ABC Technologies Newsletter* (Fall 1997).
2. Brad Ackright, "KCPL Restructures G.L. to Reflect Activities," *As Easy as ABC: ABC Technologies Newsletter* (Fall 1996). Kansas City Power and Light case is an example of doing ABM activities in parallel. KCPL split their teams into parallel developments of model creation, report design, and approvals and assignment methodology. Their process took the form of conception, data gathering, model design, reporting, and training. With the project starting in April 1995, they achieved full model operation by January 1996.
3. SAM Project Team, "Hewlett Packard Knows What It Takes and What It Costs," *As Easy as ABC: ABC Technologies Newsletter* (Summer 1995).
4. Karen Ribler and Deb Dixon, *Activity-based Management—A Primer for Foodservices Brokers* (Reston, VA: Association of Sales and Marketing Companies Foundation, 1996).

5. Dagmar Walkington, "Understanding the Cost of Post Processes and Products—The New Zealand Post Experience," *As Easy as ABC: ABC Technologies Newsletter* (Summer 1997).
6. Amanda Mcvitty, "Decision-Making Made Easy," *Management Technical Briefing* (June 1998), pp. 27–32.
7. S. Schaefer, "ABC Implementation: Planning Is Everything," *As Easy as ABC: ABC Technologies Newsletter* (Spring 1991).
8. Jim Merryman, "Food Industry Leader Uses ABC Recipe for Success," *As Easy as ABC: ABC Technologies Newsletter* (Summer 1991).
9. Christopher Dedera, "Harris Semiconductor ABC: Worldwide Implementation and Total," *Journal of Cost Management* (Spring 1996), p. 94.
10. J. Donnelly and Dave Bucanan, "US Airways Takes Off with ABC," *As Easy as ABC: ABC Technologies Newsletter* (Winter 1997).
11. John Faulds, "Innovative CityMax Turns to ABC to Define Its Future Business," *As Easy as ABC: ABC Technologies Newsletter* (Summer 1997).
12. Adapted from K. Phillips and Kevin Dilton-Hill, "Willards Foods: Managing Customer Profitability with ABC Information," *As Easy as ABC: ABC Technologies Newsletter* (Winter 1996).
13. Thomas Hoffman, "Datawarehouse, the Sequel," *ComputerWorld*, June 2, 1997, pp. 69–72.
14. Steven Covey, *The 7 Habits of Highly Effective People* (New York: Simon & Schuster, 1990).
15. Up to 15 drivers are noted to be noncorrelated.
16. Report mining systems generate asynchronous reports that are self-triggering when certain predefined data items or formulae or relationships change. Read Stewart Mckie, "Mining Your Accounting Data," *Controller Magazine* (November 1996), pp. 43–46.

11

FOURTH WAY: TREAT THE ENDEAVOR AS A PROJECT

ABC/M projects need strict and controlled sequencing. Projects tend to be overcommitted, understaffed, and miscalculated. Planning overcomes failure. This chapter provides general examples and samples of technology project management. Managing an ABC project should be looked on as managing a product development and introduction cycle rather than just project management. If project leaders view their challenge in this way, they will understand and take into account the various aspects of messaging, expectation setting, project management and communication, periodic reporting, and the formalisms of periodic deliverables and written plans that culminates in an effective introduction into the target audience. A project- and product-focused ABC/M program would:

- Assume that the needs of users will always increase.
- Define a project schedule with deliverables.
- Identify overall project phases and conceptual system design.
- Develop a detailed set of deliverables and assign owners in each phase of the project.
- Define the levels of involvement of consultants and vendors.
- Manage the activity dictionary.
- Establish a tools inventory.

NEEDS OF USERS WILL ALWAYS INCREASE

Projects must be defined and stakes placed firmly in the ground. If not, requirements will increase faster than any implementation can catch them. If the project succeeds, the expectations will increase as well. Exhibit 11.1 illus-

IDENTIFY OVERALL PROJECT SCHEDULES AND SYSTEMS DESIGN

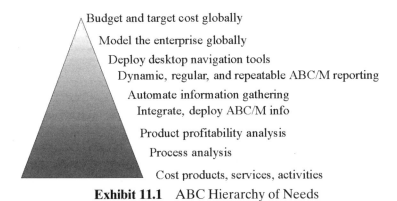

Exhibit 11.1 ABC Hierarchy of Needs

trates a hierarchy of ABC/M needs as a project matures. A survey by Atre Associates showed that many data marts were built in three to four months and grew at 100 percent or more yearly. This led Atre Associates to postulate that pilot projects, without much additional functions, become the production system.[1] Any project planning document should define the scope of the project but also highlight the depth of future implementations to ward off any challenges that tend to affect a project. This phase is very similar to the requirements phase described in Chapter 10; this document should cover more than just the model architecture. It should address various aspects of the project like resources needed, schedule issues, timelines and expected results.

DEFINE A PROJECT SCHEDULE WITH DELIVERABLES

Exhibit 11.2 illustrates a sample project schedule with timelines and deliverables. Traditional Gantt charting is appropriate. An ABC/M project could be enabled by treating it as a product introduction. Here the rigorous demands of project and product management take hold. Exhibit 11.3 drives this home with three separate views of the same project: phases, project, and document and deliverable view. Each view brings out a different yet significant view of the same project. Once a project outlines all the checks and balances in each view, it begins to form three-dimensionally.

IDENTIFY OVERALL PROJECT SCHEDULES AND SYSTEMS DESIGN

Management will not deal with complexity well. ABC/M teams must first boil down all their activities into a few key phases, as illustrated in the project worm

128 FOURTH WAY: TREAT THE ENDEAVOR AS A PROJECT

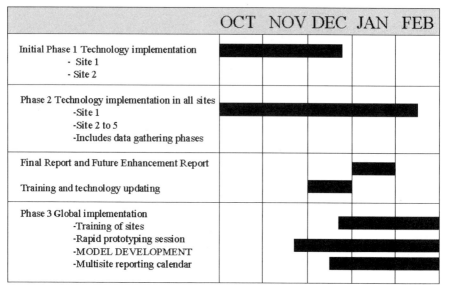

Exhibit 11.2 Example Time Lines

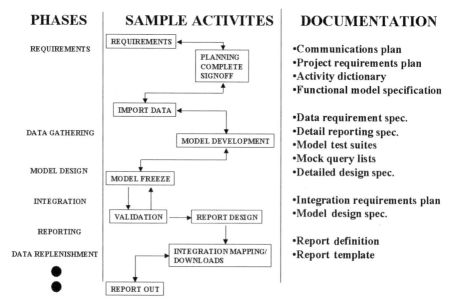

Exhibit 11.3 Views of Project

IDENTIFY OVERALL PROJECT SCHEDULES AND SYSTEMS DESIGN

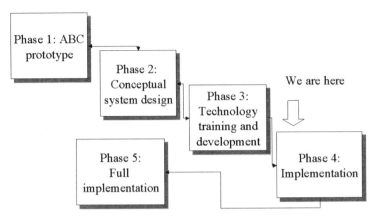

Exhibit 11.4 ABC/M Simplified Technology Project Description

(Chapter 10). Exhibit 11.4 is an example of the five basic phases that can be used to present a pilot project to management. Exhibit 11.5 is an example of a conceptual design identifying the inputs to the system, the outputs of the system, and the system repositories and data staging centers to prepare and format information. Maps like this remove many doubts on why, where, and what is needed to build a model.

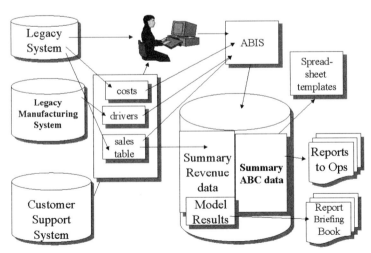

Exhibit 11.5 Conceptual Design Framework

DEVELOP A DETAILED SET OF DELIVERABLES AND ASSIGN OWNERS TO EACH PHASE OF THE PROJECT

Exhibit 10.7 illustrates sample deliverables in each phase of the pilot. These outputs are critical to the success of the project. Minimilestones within each project phase will alleviate the sense of hopelessness that overcomes any project team as they view the project as a whole.

DEFINE THE LEVEL OF INVOLVEMENT FOR EACH CONSULTANT AND VENDOR

ABC/M projects can be complex challenges if the roles of consultants, vendors, and in-house teams are not defined. Up front, successful program leaders identify the required functions for each of the players and then manage their effectiveness. Exhibit 11.6 illustrates such a mapping for a project. Once this responsibility chart is drawn, any resource mix-ups or role misunderstandings will be identified. Also, resource problems will be brought out at the beginning of the process. Some ABC/M projects have lasted so long that the staff on the project leaves or changes. It is very important to know the level of contribution and participation demanded of each member so that resources can be brought in to fill the need.

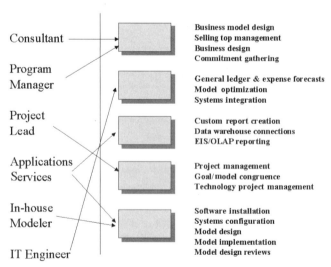

Exhibit 11.6   Types of Work in ABC/M Project

MANAGE THE ACTIVITY DICTIONARY

Many ABC/M projects end before they begin. While these projects use all the sophisticated technology, strong management support, and project management, they still stumble.

The secret ingredient seems less process-oriented but content-specific. The use of an activity dictionary is critical to establishing the language of the project. In Chapter 10, activity dictionaries were discussed as one way to start the design process. Exhibit 10.6 illustrates a sample page in a dictionary. Note that it describes an activity, its members, its process parents, inputs, cost drivers, and performance measures among others. Further additions include:

- The nature of drivers—batch, unit, sustaining or product
- The level of the activity—primary or secondary
- The nature of the attributes
- When and how to collect its data

An activity dictionary isolates the more important contents of the model prior to beginning the project. Although everyone sitting in a room defining activities may seem to compromise the momentum of the project, the value of this exercise will prove itself later in the project.

ESTABISH A TOOLS INVENTORY

Just as a design schematic clarifies sources and uses of data, a tools schematic topologically identifies the necessary tools/solutions for the project. Exhibit 11.7

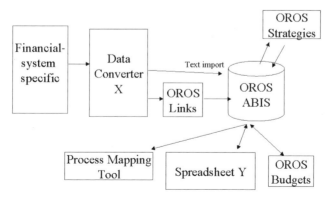

Exhibit 11.7 Conceptual Tools Design

illustrates such a topological view of tools. Other tools also can assist in the project and safeguard technology crises that may surface. If alternatives are identified, the project must ensure that data models can be converted prior to the development start.

PRODUCT AND PROJECT PHILOSOPHY

Certain firms actually have sent their ABC/M team members to their own in-house project management classes prior to starting an ABC/M project. Others have insisted on strict product life-cycle guidelines, assuming that the release of the ABC/M model and resultant information is like a product release. These two methods have found their way into the mainstream success methods.

NOTE

1. Shaku Atre, "Plan for Data Marts," *Computerworld,* June 16, 1997, pp. 71–72.

12

FIFTH WAY: WATCH THE EIGHT OBSTACLES

Often ABC/M project learn from the "failures" of others. However, learning from success is very possible with ABC/M projects because examples now exist. As one of Mexico's leading wholesale distributors with 1997 revenues of 8,746 million pesos (more than $1 billion U.S.), Grupo Casa Autrey is just such a success.[1] "With a clear idea of the business, use of focused software and a smart corporate strategy, you can model the functionality of the enterprise," states Jorge Medina More E., who interviewed Ruben G. Camiro, CFO of Grupo Casa Autrey.

Camiro declares: "In the technical implementation there were obstacles to overcome, interfaces that needed to be generated, and other technical problems to be solved. These problems were anticipated, but they were few and workable."

OBSTACLES TO SUCCESS

Organizations that succeed in their implementations anticipate and overcome the following obstacles:

- Data-gathering time: a technology challenge
- Education of users: a people challenge
- Management understanding and support: a people challenge
- System maintenance and data replenishment: a technological challenge
- "Freeloaders" resist change: a people challenge
- Searching for "push-button" solutions: a technological challenge
- Expecting the model to freeze: a technology challenge
- Hidden costs: a resource challenge

Several of these challenges have been addressed in prior chapters, but it is important to stress the simplicity of solutions if problems are anticipated and overcome even before they arise.

DATA-GATHERING TIME

Data-gathering and data-cleansing demands are challenging, but organizations do not suffer from too little data. Someone, somewhere, somehow in the organization at any time is collecting point-oriented data. It is easy to be saddled with too much data. The skill is in identifying the correct sources of data and drivers.

EDUCATING USERS

Educating users is an unending challenge. Other chapters have discussed the value of educating users and have suggested ways to do so. Knowledge workers who understand the value and use of ABC/M will make decisions using an activity-based information system. Others will mistake ABC projects for another irrelevant tool. While conceptual training in the value and use of ABC is essential, technical training is also important. Training needs to take place in the following areas.

- ABC/M fundamentals. This includes learning the basics about ABC, the CAM-I modeling method, and the output measure approach, and understanding a case or two and how to approach the challenges of change management.
- Modeling basics. This area involves knowing how to design and build a basic model using the software; learning the CAM-I methodology to get started; and connecting resources, activities, and cost objects to create a basic model. (See Exhibits 12.1, 12.2, and 12.3)
- Case studies. Learning from others' mistakes and successes is key to building an ABC/M program. Looking at cases in the industry being considered will bring a whole new light to the project. Not all industry implementations are similar, and matching objectives to models is key.
- Integrating data sources with an ABIS. Almost 70 percent of the model can be imported from various sources. Cases exist of models built with 30 different data input sources that need to be automated. Both ASCII importing and query-based importing must be examined. (See Exhibits 12.4 and 12.5)

EDUCATING USERS

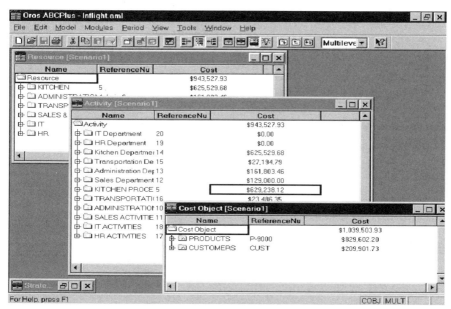

Exhibit 12.1 Modeling Resources, Activities, and Cost Objectives (Reprinted with permission of ABC Technologies.)

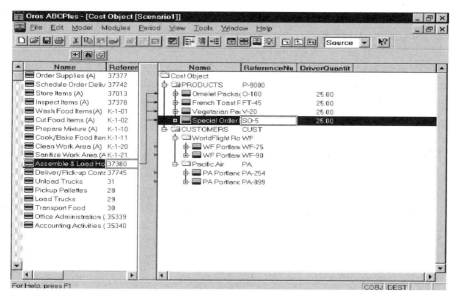

Exhibit 12.2 Making Assignments from Activities to Products and Services (Reprinted with permission of ABC Technologies.)

FIFTH WAY: WATCH THE EIGHT OBSTACLES

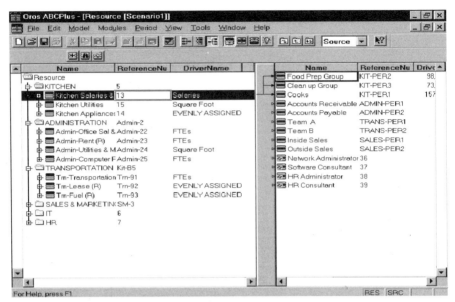

Exhibit 12.3 Making Assignment Paths from Resources to Activities (Reprinted with permission of ABC Technologies.)

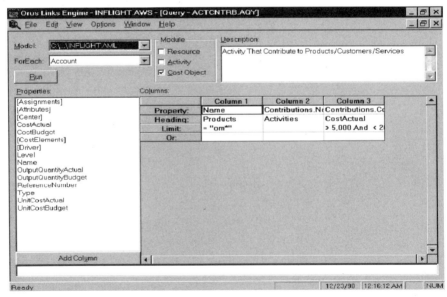

Exhibit 12.4 Query Engine Interrogating ABC Model File (Reprinted with permission of ABC Technologies.)

EDUCATING USERS

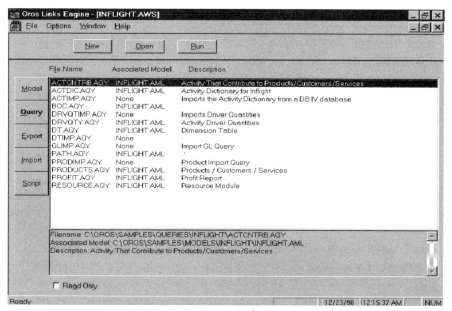

Exhibit 12.5 Database Query Engine (Reprinted with permission of ABC Technologies.)

- Profitability analysis skills using technology. This section is applicable to organizations that strive for profit, which does not include nonprofits and government entities. For many organizations, a major part of the modeling need is to know how profitable, using ABC, their products, services, channels, partners, and vendors are. Organizations also want to know how to mix and match these different views multidimensionally (i.e., ask questions like "What is my cost or profitability by this channel for this product with this vendor?") (See Exhibit 12.6.)
- Optimizing and designing a model. Many times models are built with very little thought as to which is the best and most optimized way to work with the software. Consequently, when it finally comes down to running the model, calculation times may take 24 hours to complete due to the model size. Models cannot be measured by size; they should be measured by output and by their scope.
- Data mapping and cleansing. Many projects underestimate the tedious work required when multiple data streams are brought into one model. Unfortunately, the many silos of data store data in different syntax and formats. Data cleansing requires repetitive work and skill. Data mapping is the art of linking all this data into one common model.

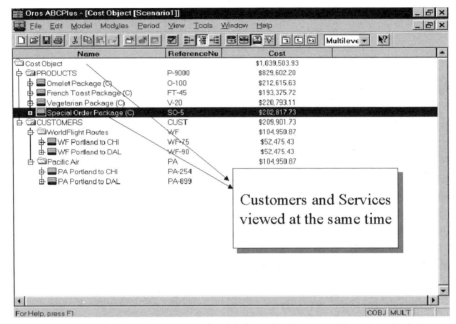

Exhibit 12.6 Viewing Customers by Airlines by Routes (Reprinted with permission of ABC Technologies.)

- Collecting empirical information. The trouble with collecting information is that it is not as exciting the second and third time. Collecting information found in people's heads about process and the work being done is even more challenging. Technology can help in this process. Survey-based input software is available. (See Exhibit 12.7.)
- Mastering the use of attributes and drivers. One of the secret weapons found in a model is the use of attributes to tag any item. For example, an activity can be tagged with a value attribute choosing between "non–value-added" and "value-added." Then the model can report on any costs or data element related to these value attributes. Activity drivers (e.g., number of sales orders) give the user an understanding of what drives their activity and what products or services the activity consumes. By understanding drivers, users can affect the performance of an activity using these drivers. (See Exhibit 12.8.)
- Sustaining models. The first time around, every ABC/M exercise seems exciting, and it would be easy to cut corners and get to the final result. The next time around, and the time after that, careful attention to detail and knowledge to sustaining models is required.
- Multimodel management. Consider having 10 models in 10 different

EDUCATING USERS

Exhibit 12.7 Electronic Data Replenishment with Surveys (Reprinted with permission of ABC Technologies.)

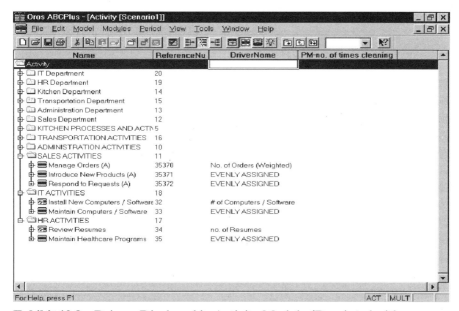

Exhibit 12.8 Drivers Displayed in Activity Module (Reprinted with permission of ABC Technologies.)

countries whose results are reported on. Technology and formalisms exist to support this topology.
- Consolidating models. Many times many models in different places need to be consolidated and combined to give a total picture of the enterprise.
- Designing an executive information system. Flat reports are fine but technology now exists that can pour a model into a cube so that users can "slice and dice" the information. ABC/M data lends itself to this very well because all the information contains hierarchy and cause and effect, and has depth.
- Process analysis and yield analysis. Process-sensitive industries and industries that rely on a process or recipe to produce products and services, such as semiconductors, food, or paper, must have an understanding of their step yield and their cost per process element. Using activity-based information, the yield of each step is more accurate and essential.
- Designing custom report. Reports still top the list of outputs for an ABC model regardless of all the other ways to interrogate the model. Several reports are unique to an activity-based model. For example, a contribution report shows the user what contributes to each product or service, moving from products and/or services up to activities and to resources.
- Flowcharting. Even though ABC is about cost flow, processes and activities are about time-based flow. In some ways, costs flow from top to bottom while time flows from left to right. They meet at the activity or process level. (See Exhibit 12.9.) Flowcharting is an essential reporting output as well as an analytic tool to investigate the way work is performed.
- Linking ABC to scorecards. The introduction of Balanced Scorecard methodology has opened up the world of performance measurement.[2] ABC and scorecards or performance measurement systems feed each other. The technology now exists to link both of these functions together so that the cost and profitability elements of ABC/M can add to the vast performance measurement schemes established by organizations.

MANAGEMENT UNDERSTANDING AND SUPPORT

Strategy is in the hands of the management while implementation is in the hands of the operating teams. Management's attention is highly elusive and must be focused for an ABC/M project to succeed. ABC/M must identify "pain" and exploit it to win. ABC/M projects must be tied to strategy and must

SYSTEM MAINTENANCE AND DATA REPLENISHMENT

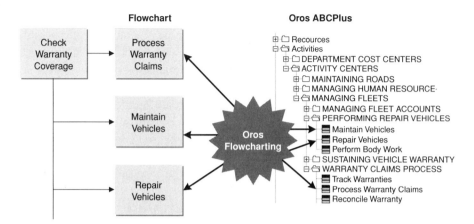

Exhibit 12.9 Flowcharting and ABC (Reprinted with permission of ABC Technologies.)

take advantage of the economies of scale using other initiatives to bargain for resources. Strategy, key performance indicators, activities and ABC should be closely associated in an ABC/M endeavor.

One method to secure management involvement is to use a "rapid prototyping" method to permit hands-on understanding. Picture two main players in a room of executives—one facilitator and one keyboardist. One gathers information and sets the playing field, bringing out the best in the audience, while the other records the input into the information system software. At the end of the highly interactive exercise, a high-level management model is built to illustrate the strategic value of ABC/M. Outputs are reviewed and analyzed preliminarily. These methods have "charged-up" management teams around the world and gained their immediate commitment even before the process has begun. In some ways, using software like this gives managers the impression that "what you see is what you want."

SYSTEM MAINTENANCE AND DATA REPLENISHMENT

It is important to gain the involvement and respect of information technologists early on the ABC/M cycle. Without their support, the project will remain a pilot, off-line endeavor. As mentioned, the true enterprise systems are finance-endorsed, operationally used, and IT maintained. Due to limited resources devoted to IT, combined with the overwhelming demands of the challenges of Y2K, most IT organizations will resist any new endeavor. Once they understand and respect the strategic nature of ABC/M, their support will follow. Several pilot programs used their own finance resources and did not bother the IT personnel until proof-of-concept was completed. Other implementations got resources allocated early in the cycle and devoted them to the multifunction teams that drove ABC/M in the organization. Both methods work and have proven out well. Nevertheless, system maintenance is critical and usually underestimated.

"FREELOADERS" RESIST CHANGE

People who enjoy the benefits of distorted information tend not to want change. Why should they confess to a change when they have found themselves on the top of the value chain? In any organization, there are perceived winners and real winners. Real winners always look for real answers; perceived winners prefer the status quo even when on the *Titanic*. Management endorsement and apolitical encouragement will overcome these barriers, but until then, everyone wants expenses to belong to someone else. The organization that resists any change should worry the champion. Without management support and a logical organizational culture, the likelihood of success should be questioned.

SEARCHING FOR PUSH-BUTTON SOLUTIONS

Humans have expected "push-button" solutions ever since they touched a computer. Such solutions will happen only if change never occurs. The nature of an ABC/M project is change—when we use ABC, we will change the structure of our activities and the data will change periodically. Hence push-button solutions tend not to be characteristic of an ABIS.

EXPECTING THE MODEL TO FREEZE

Chapter 10 elaborated on model design and emphasized the importance of building flexible, expandable models. Yet no one can anticipate the transformations that come about after results are shared. Thus, ABC/M practitioners may expect to redesign and rebuild their creations regularly. This is a good problem to have. Organizations that consistently repopulate their models with data and report on them are on target. The real reason ABC exists is to change not just the data but the structure of business and, hence, the structure of activities. Only then will related improvements surface. If no changes occur, then no one is improving.

HIDDEN COSTS

Hidden costs will kill the continuity of a project. Consider the following costs in anticipating the resource requests of the project:

- Training and retraining costs
- Educational tools (books and videos)
- Consultant costs with possible overruns
- Software upgrades and updates
- User group attendance
- Visits to vendors
- Emergency calls and trips
- Integration technology costs
- Support schemes (number of calls before the charges begin)

Activity-based information systems are not complicated or confusing if they are designed and developed with these eight obstacles in mind. Successful ABC/M projects often forget the enormous number of obstacles that appear throughout the journey. Yet if problems are anticipated and alleviated, a project can be smooth and uneventful. ABC conference attendees will attest that the challenges are finite and definable. But the solutions to them are organization specific. In other words, solutions that work for one organization may or may not work in another.

However, technology issues seem to be definable and predictable if the ABIS is understood and the relevant training, consulting, and applications services are deployed and bounded. Many times the software technology capabilities are assumed until the last moment. Not all software is created equal; the systemic problems in many projects are related to models being designed as they go along. This methodology is dangerous but commonly used.

The fifth way reminds project leaders to anticipate, take the ABC/M challenge in stages, and realize that the project will fail if problems are not anticipated and worked around. Above all, projects that are defined will have the best chance to succeed.

NOTES

1. Jorge Medina More E., "Grupo Casa Autrey's CFO drives profitability using ABC," *As Easy as ABC: ABC Technologies Newsletter* (Summer 1998).
2. Robert S. Kaplan and David P. Norton, *The Balanced Scorecard* (Boston: Harvard Business School Press, 1996).

13

SIXTH WAY: ALIGN WITH STRATEGY

The ABC world is populated by three kinds of personalities:

1. Those who adopt ideas and technology early and take the steps before others do.
2. Those who follow those who take the first steps after the early adopters have tested it.
3. Those who never take any steps.

All successes in the ABC/M world are due to the first two personalities. The early adopters have reaped the rewards and the problems in this market for the last 10-plus years. The only way to finish any technology implementation is to start.

But the ABC/M journey is never completed. That is actually a good sign. This unending nature of analysis using ABC information identifies it as more than a fad; rather it is an instrument or a compass to direct the future of organizations that attempt to understand, experiment, and command the information.

Many organizations have asked if ABC will ever get off the ground while others have been making decisions using advanced ABC information for over 10 years. The organizations that have seen strategic value in ABC tend to share their success less and less. This natural and expected behavior sometimes makes ABC/M on of the best kept management secrets.

Meanwhile the early adopters have transformed the culture of their organizations from operationally managing through the rigid bars of traditional accounting to activity-based decision making. The true measure of strategy is more than just execution or market acceptance; it is the ability and readiness for change. If organizations debate the value of water pumps when a ship is sinking, no wonder the ship sinks. Lifeboats come in all shapes and sizes, but the difference between a successful crew and another is the displayed ability to make decisions, right or wrong, appropriate or not, and to change the decision once experimented with.

UNDERSTANDING STRATEGY AND ACTIVITIES

When Michael Porter spoke at the seventh annual ABC Technologies International User Group conference in Chicago in 1997, he encouraged his ABC/M audience to march on in their mission. He spoke of the importance of linking activities to strategy and said that strategic fit of activities was a profound measure of strategy. In a dialog with Porter, one high-ranking participant stated that he used strategy and ABC/M to align his activities. He claimed that if any activity was not aligned to a strategy, it would either be removed or moved to a lower level in the hierarchy of activities. This method serves two purposes:

1. As a way to determine the value of any activity using strategy
2. As a way to monitor the size and hierarchy of a model so that it does not get too large and complicated

According to Porter, strategy has been incorrectly viewed in the context of operational effectiveness. Using the language of activities, he has outlined a way to differentiate between strategic positioning and operational effectiveness.[1] Operational effectiveness is performing similar activities better than rivals, while strategic positioning is performing different activities or performing similar activities differently. In some ways, when organizations do anything as a competitive advantage using operational effectiveness, others follow. W. Chan Kim, of the Boston Consulting Group, Bruce D. Henderson, professor of international management at INSEAD in Fontainbleau, France, and Renee Mauborgne, a senior research fellow at INSEAD, put it well when they said, "The trouble with forging a highway is that if you are right, imitators will follow. Then you are back into protecting your base and become subject to conventional wisdom."[2] Porter emphasizes that benchmarking only makes companies similar. Just using ABC/M to identify and improve the activities in a company or improve the cost management aspects of the business does not forge a competitive advantage. However, using ABC/M to align activities to strategy offers tremendous value.

Strategy: The Messiah Method

Moses went to the mountain, meditated, found enlightenment, and returned with the 10 Commandments. Strategy sessions in many organizations happen

UNDERSTANDING STRATEGY AND ACTIVITIES

the same way, with executives locking themselves in conference rooms or resort hotel rooms to find their "tablets." But strategy formulated with no regard to strength and weaknesses in capability is blind. The true power of strategy is illustrated only in work performed. Furthermore, the business world is guided by change, and it seems that change can affect business models drastically. Mergers and acquisitions can transform the competitive landscape as power shifts. Hence, sticking to a good-looking strategy when strategic variables change can be dangerous. For companies to be effective, they must have the capability to adjust to a change in strategy more than the ability to formulate a strategy. They must have the ability to:

- Formulate strategic thrusts, that is, several key areas of focus and strength.
- Institutionalize and operationalize strategy into key performance indicators (KPIs) or sets of activities that if performed would enable the key strategy.
- Manage resources that trigger key activities resulting in the expected performance.

Strategy without strategic alignment to KPIs and activities renders organizations impotent. The strength of a competitive organization comes from its ability to change its strategic thrust and see it reflected in actions and corresponding performance measures. This connection among strategy, KPIs, and activities can be achieved using ABC/M. Aligning an ABC/M program to these higher goals will strengthen and focus a mission and align it to the strategic program on the desk of decision makers. For example, consider a high-technology chip company that competes in the fast-paced semiconductor manufacturing business. This company may have the following strategic measures:

- Strategy
  —Dominate, with 60 percent share, the semiconductor market targeting the high-end computer "server" market
- Strategic thrusts/KPIs

1. Be the market supplier of choice to server providers.
2. Establish and dominate in customer service.
3. Align with Intel Corporation by providing peripheral chips on their single-board computers.

- Activities serving strategic thrust or KPIs are:

1. (a1, a2, a4)
2. (a4, a5, a8)
3. (a4, a9, a7) where a = key activity such as "develop chip technology at lowest cost"

Exhibit 13.1 illustrates such a strategy with emphasis on order of priority. Michael Tracy and Fred Wiersema, authors of *Discipline of Market Leaders*, declare three strategic thrusts to market leaders to be:

1. Operational excellence
2. Product leadership
3. Customer intimacy[3]

If optimized, various sets of activities can make each one of these thrusts achievable. Using ABC/M, activities are aligned with resources, products, and services provided by the firm. With this ABC analysis, these assignments would provide more information about the cost and effectiveness of the activities. Meanwhile, management will establish the priority of the strategic thrusts as shown in the exhibit, recognizing that the ABC/M analysis will give a measurement to these thrusts soon.

With good bottom-up analysis, the ABC/M project could specify the emphasis of the resources on key activity sets (KPIs). This analysis shows a different emphasis when compared to what is truly performed (shown in dashed lines in Exhibit 13.2). Comparison between resource really placed on activities can be compared against what work is thought to be emphasized. The result is called a paradox map. Since the charts are based on activity-based information, users can drill down on the map to isolate what is really going on with the activities

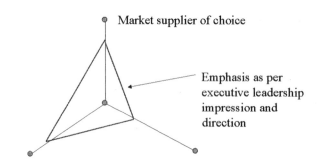

Exhibit 13.1 Strategic Thrusts

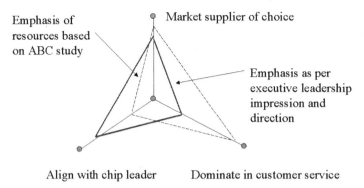

Exhibit 13.2 Paradox Map

that serve a particular strategic thrust. (See Exhibit 13.3.) Comparing any two variables, users can compare how the activities or sets of activities align. Actions can be taken to improve the measured ratios (in Exhibit 13.3, the ratio is cost to value) taking one activity at a time or one activity set at a time.

In this exhibit, the ABC/M project has identified that the priority of resource emphasis has been in another order. Hence organizations can carve their competitive advantage, drill down to see what resources can be adjusted, and even see the costs transform using this structure. ABC/M brings the power of strategic alignment to the forefront.

Imagine what a problem a CEO could cause if he or she forced a change in strategy emphasis a second time, unaware of such misalignments.

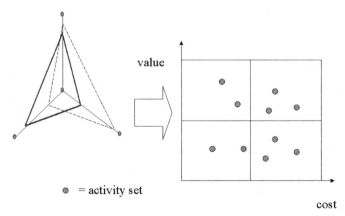

Exhibit 13.3 Drill Down to Elements

STRATEGY AND ABC/M

The number-one reason why ABC/M projects fail is the lack of senior management support. Senior management dollars and coaching drive the success and endurance of any ABC/M project. If senior managers knew of the relationship between strategy and ABC/M, this support would be easier to achieve. Costs are for cost managers; efficiency is for operational management; strategy is for executives. If the ABC/M project goals do not align with the strategic emphasis of the firm, the project may succeed but the organization may not. Boardroom expectations, operational goals, and ABC/M objectives must align and coincide for an ABC/M project to sustain itself. Exhibit 13.4 illustrates a model that links ABC to strategy. ABC/M projects that succeed provide:

- An analysis of the impacts of different strategic thrusts.
- An analysis of the connections and relationships between:
  —Strategy and profitability
  —Strategy and resource consumption
  —Strategy and activity costs
  —Strategy and value-added analysis

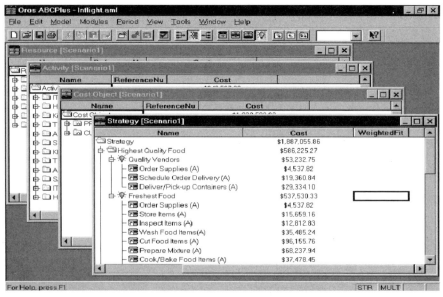

Exhibit 13.4 Linkage of ABC to Strategy (Reprinted with permission of ABC Technologies).

Strategy and ABC/M are related just as upper management priorities are related to the work being performed. Tools are emerging to assist in this relationship. In the end of the ABC/M mission, the results should provide insight into the true workings of strategy. As Craig Weatherup of Pepsi claimed, eventually strategy leads to processes because "capability comes only by combining a competence with a reliable process."[4]

NOTES

1. Michael Porter, "What Is Strategy," *Harvard Business Review* (November–December 1996), pg 61–78.
2. W. Chan Kim and Renee Mauborgne, "Value Innovation: The Strategic Logic of High Growth," *Harvard Business Review* (January–February 1997), p. 106.
3. Michael Tracy and Fred Wiersema, *The Discipline of Market Leaders* (Reading, MA: Addison-Wesley, 1995).
4. D. Garvin, "Interview with Craig Weatherup of Pepsi: Leveraging Processes for Strategic Advantage," *Harvard Business Review* (September–October 1995), pg 76–90.

14

SEVENTH WAY: PLAN FOR ENTERPRISE-WIDE EXPANSION

ABC/M projects evolve from small pilot programs to enterprise systems, local or global, and eventually use ERP systems in concert with analytic ABC/M applications.

Several pilot programs have turned to enterprise-wide deployment with careful project management, technology deployment, and model design consistency. They have succeeded. Mobil Oil's global ABC implementation, a two-year project with 45 implementations, has shown significant results identifying even a 20-fold variation in cost among products.[1] Prior to ABC, some products had reported distorted production costs.

Chapter 6 described enterprise-wide systems. Clearly, analytic applications—that is, stand-alone ABC systems—cannot claim to be enterprise-wide production systems, because they lack enterprise scalability. However, they are the root of modeling, prototyping, and learning and are the precursor to any enterprise-wide ABC/M implementation. ERP systems serve the enterprise reporting and control functions and are the root of operationalizing ABC/M across the enterprise. ERP and analytic systems need each other to be an activity-based information system. Together, the analytic systems and the operational system form the enterprise systems. Prior to creating an ABIS architecture, analytic applications can lead in the learning and the prototyping. The chapter discusses this phase of the project where the organization moves from pilot to the next phase of experimentation, which is to model the enterprise usage of ABC/M.

Threats to project implementation exist at every transition, from pilot to production or from production to global distribution. As in a relay race, the race is lost not in the running but in the passing of the baton to the next runner. Expanding the enterprise-wide system implementation has similar hand-off challenges. A few critical hints against such hindrances are:

- Design the team for enterprise expansion.
- Expand the model control technology.
- Build the enterprise configuration.

DESIGN THE TEAM FOR ENTERPRISE EXPANSION

Organizations walk a fine line between absolute control and absolute freedom in managing and implementing global sites. Is the philosophy centralized or decentralized, or (probably) something in between? What does the organization want as a standard? If this standard is insisted on, would the organization lose the true value of diversity and creativity? Other possible issues are:

- What interfaces does the organization use in technically automating the process?
- What level of integration is expected in various sites, and do they have the same technical capabilities?
- Are the knowledge and skill levels up to par? Do they vary from site to site?
- How do different sites change their models or increase their expectations? Is there a "change process" that must be approved by the central controlling body?
- Are all implementations at the same level?

Steve Player of Arthur Andersen identifies the following best practices:

- Start with a cross-country steering team.
- Use a multilingual newsletter.
- Draw from global best practices.
- Use a multicultural, multilingual consulting team.
- Develop multilingual training materials.
- Use standard activity dictionaries with a tiered structure.
- Deploy an internal company-wide user group.[2]

These are great hints to enable the process. One that demands attention is the activity dictionary and its value in the global rollout. Once again, a tiered approach in a dictionary would allow for site variations as well as maintaining core activities without redundancy and definitional errors. One large semiconductor firm uses a dictionary of this nature; it includes a brief description of the history of ABC and its value, implementation hints, and an adjusted, customized version of the CAM-I cross to bring readers and implementers closer to their excitement.

Many organizations that wish to expand beyond their local sites to multiple ABC endeavors worldwide form internal competency centers to dispense and distribute learning and technology. These centers are responsible for:

- Educating and training users
- Ensuring design and implementation consistencies
- Central technical support
- Software selection and negotiation
- Functioning as a clearinghouse for upgrades and updates of software
- Certification of model architectures and model consistencies

EXPAND THE MODEL CONTROL TECHNOLOGY

Enterprise deployment objectives usually include two goals:

1. Model consolidation
   - Model globalization
   - Roll-up of tactical models into a strategic model
   - Year-to-date consolidation from several single-period models
2. Modeling to corporate standards
   - Checking compliance to allow consistent reporting
   - Establishing template models

Inherently, most global-enterprise rollouts should consider the following:

- Local organizations resist tight control of model creation.
- Complete control of local models is unnecessary and removes local cultural contribution.
- Compliance checking must be simple and painless for the local modeler.
- Standards can be easily created using common and understandable tools.

Available technology allows for consolidating models and establishing standards. In the case of model consolidation, technology now enables ABC/M model elements to be linked together between models. Model linking, as it is called, links accounts across models and simplifies model consolidation considerably. In the past, model data had to be exported and imported into new models, increasing the risk of errors.

Model certification and verification ensures that models created in various sites live within certain communicated and agreed-on guidelines. Exhibit 14.1 highlights the hierarchy of "rules" that can be placed on models. Rules tech-

DESIGN THE ENTERPRISE CONFIGURATION

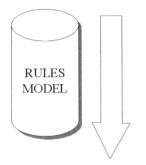

- Standard structure compare
- Enforce naming convention
- Enforce/verify usage of attributes/drivers and bill-of-costs
- Assignment matching
- Model-wide rule enforcement
- Importing/exporting rules

Exhibit 14.1 Model Rules

nology permits model designers to formulate a set of model criteria and structure to be used as a guide or template to check. It also enables them to:

- Establish a base model that contains all rules.
- Export rules from one rules model into another. (See Exhibit 14.2.)
- Interactively check a model against established standards as it is being built.
- Verify actual objects and their positions with the model.
- Compare models to isolate differences.

DESIGN THE ENTERPRISE CONFIGURATION

Three deployment architectures are common in enterprise deployment:

1. Hub-spoke
2. Hierarchical
3. Hybrid

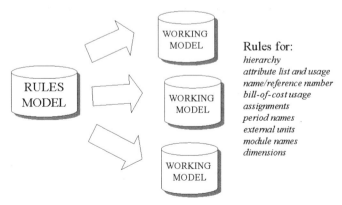

Rules for:
hierarchy
attribute list and usage
name/reference number
bill-of-cost usage
assignments
period names
external units
module names
dimensions

Exhibit 14.2 Model Certification

In hub-spoke, ABC/M projects would expand from a central core team into global locations. Hierarchical projects are developed in a top-down fashion with models being decomposed into feeder models. The hybrid approach combines both these configurations. Exhibit 14.3 illustrates all these methods. In (a) control, training, education, certification would rest on a central hub while in (b) they are shared responsibilities. None of the architectures is inherently defective. Choices depend on the culture of the global entities in question and the process rather than technological factors.

NOT ALL WORLDS ARE CREATED EQUAL

In considering deployment and remote model development, the natural expectation is to bring models up to the same level at the same time. Doing so would place a heavy burden on the ABC/M project champion as well as on the learning and development curve of every team across the globe.

Sequencing all aspects of these models may be essential to ultimate success. For example, in the areas of data gathering, model results/reporting, model design and structure, training levels, and knowledge transfer capabilities, consider expecting different levels of capability from each site. Not all sites can be at the same level, given different resources. Exhibit 14.4 illustrates such a program where A, B, and C are at peak levels while D, E, and F are provided a longer runway initially. Enterprise projects are always training and developing new techniques at their own pace. Hence, walking cautiously and stepping up expectations and skill development is essential.[3]

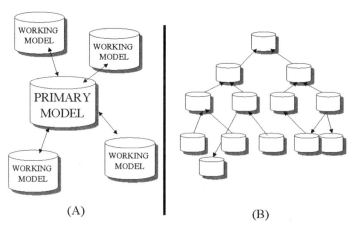

Exhibit 14.3   Expansion Architectures

	A	B	C	D	E	F
Data collection	2	1	1	5	7	9
Model structure	1	3	4	2	1	6
Integration	4	3	1	4	4	5
Reporting	3	3	1	1	1	1

1=HIGH
4=LOW

Exhibit 14.4 Leveling Y Sites

Enterprise deployment and use of ABC information is on the rise. Systems are being developed to handle all the growing requirements of such an exercise. Technology is as critical as the education and training necessary to use ABC information at the operational level. Only then can organizations claim that ABM is under way.

NOTES

1. Henry Morris, "ABC Technologies: A Business Methodology Foundation for Analytic Applications," *IDC Bulletin,* No. 16838 (August 1998).
2. Steve Player, "Seven Key Project Aids for Global Services," *Cost Management Trends* (October 1997).
3. Ashok Vagama implemented this idea in his own organization. He spoke of this at the December 1998 CAM-I conference.

15

LESSONS IN SUCCESS

More examples exist of lessons learned in doing an ABC project than successes. Admittedly, the ABC/M business is still in its infancy. Supply chain management will find ABC a healthy preexercise before any other analysis; ECR (efficient consumer response) has found this to be the case as well; business process reengineering could benefit from the focus that is derived from ABC information.

The following four case studies are designed to show the value of ABC/M programs.

1. Case 1: Willard Foods: Managing Customer Profitability with ABC Information[1]
2. Case 2: Providence Portland Medical Center Gets a $1.6m Shot in the Arm[2]
3. Case 3: US Airways Takes Off with ABC[3]
4. Case 4: Grupo Casa Autrey's CFO Drives Profitability Using ABC[4]

CASE 1:
Willards Foods:
Managing Customer Profitability with ABC Information

By Keith Phillips, National Brands Ltd., and Kevin Dillon-Hill, World Class International

INTRODUCTION

In businesses whose strategic thrust is to be customer focused, the accounting function's cornerstone for adding value is providing customer profitability information determined on Activity-Based Costing (ABC) principles. Customer profitability information helps to decide what to do, with what product, for which customer. It translates strategy into action by answering the following types of questions:

- Do we push for volume or for margin with this customer?
- Is there scope to change the way we package, sell, deliver, etc., to improve this customer's profitability?
- Does the turnover justify the discount/promotion structure we give this customer?

Willards Foods, a division of National Brands Ltd., a subsidiary of Anglovaal Industries Ltd., holds approximately 34 percent of the South African snack food market. Approximately 90 products are manufactured at two sites and are distributed to 24,000 outlets via 25 depots across 200 routes. This article describes what Willards Foods did to provide customer profitability and process cost and performance information to better manage its business.

ISSUES FACING WILLARDS FOODS

Porter's value chain [M. E. Porter, *Competitive Advantage* [New York: Free Press, 1985]] was used as the basis for defining Willards' primary and support processes. Each process was then defined in greater detail by identifying up to 10 activities. The processes that constitute the primary (product touching) processes are:

- Inbound Logistics—The procurement and storage of raw and packaging materials
- Operations—The issue of materials into the factory, product manufacture, packing and sealing of boxes
- Marketing—All advertising, marketing, and promotional activities
- Outbound logistics—Centralized warehousing, freight to depots, and depot warehousing
- Sales and Distribution—Taking orders, delivery from depot to customer, invoicing, and collection of payments

Inbound Logistics, Operations, and Marketing are product-related costs that vary primarily by the nature of the product and the pack size. Outbound Logistics and Sales and Distribution are customer-related costs that may vary primarily by the location of the customer outlet and the nature of the Sales and Distribution process.

APPROACH TO DEVELOPING BRAND/CUSTOMER PROFITABILITY

The project to develop customer and brand profitability was split into two phases: (1) a pilot project to prove the concept and develop an implementation plan and (2) the actual implementation.

Pilot ABC Project

A pilot ABC project initially was implemented to prove the value of the information and gain executive buy-in. A snapshot of six months of actual data was used to illustrate the information that could be produced by an ABC approach to customer profitability. The following steps were taken to implement the pilot:

- Value chain processes were identified.
- Key activities within each process of the chain were defined, including appropriate output measures and performance measures.
- Five project teams comprised of different functional representatives were established to collect financial and nonfinancial data.
- The data collected by each team was used in spreadsheets to illustrate and test basic ABC principles for each process. The results of each process were consolidated into a "snapshot" result for a sample of products and customers.
- Mock-up customer and product profitability statements were developed.
- The full-scale implementation approach and project plan were developed.
- The findings and recommendations were presented to the executive.

Pilot Project Output

The pilot project highlighted several issues for the business. Previously, a key measure was the "cost per case," where various summary costs were divided by the number of cases produced. The ABC model proved that this was a gross simplification of the business and was more likely to mislead than to provide insight. Because high labor and capital costs were incurred in establishing a three-shift manufacturing environment, nonworking time in Operations had a significant effect on product cost.

The sample ABC customer profitability information clearly demonstrated that outlet profitability would allow Willards to reengineer their trading relationships with their customers to manage profitability. However, because of the cyclical nature of the business, the snapshot data was not good enough to illustrate the dynamics of the customer relationship: monthly information was necessary. Summarizing activity-level ABC costs up to their processes also helped to identify the future directions for ABC development.

It was obvious that although the In-bound Logistics and the Marketing processes managed large cash outflows, their process costs were insignificant relative to the potential costs if their activities were performed ineffectively. Consequently Willards decided to focus their process management efforts on process effectiveness and control the costs using good old-fashioned budgets.

As far as the Operations process was concerned, the key issue was line utilization. The capacity for each machine in the line was determined and the bottleneck machine that was the throughput constraint was identified. However, for the initial implementation we decided to split each production line only into processing and packaging because their different cost drivers (kilograms and packs per case respectively) yielded good insights into cost behavior in the different parts of the manufacturing process. At a later stage the process costs would be broken down into greater detail for better management and improvement of the individual operations.

The Outbound Logistics ABC analysis lead to a questioning of the extent to which cost savings would be achieved by moving to "roll-on; roll-off" technology. This process was simplified for the initial implementation by calculating an average cost to move a case to a depot. The ABC analysis of the Sales and Distribution process highlighted the cost differentials between "pre-selling" and "knock and drop" routes and the dynamics of route profitability. Willards is in the process of implementing hand-held computer technology for route accounting. The detailed data recorded by route accounting provides the ideal input into a route ABC system. It will allow us to identify each call that is made, how long the call took, and the product quantity sold. A key factor influencing route profitability is the number of calls that do not yield reasonable sales levels. However, for the initial implementation, this process was simplified by averaging the route costs per client on the route.

The rationale behind simplifying the data in the ABC implementation is to ensure that a successful implementation can be made in a three-month window. It is better to have information of reasonable quality quickly than to wait 18 months for perfection with a high risk of project failure. After the initial implementation, the model can be extended down into the detail in a controlled fashion—also in three-month windows.

Focusing on key result areas using the information developed by the pilot has already identified an annual savings in excess of R6 million to the division—many times the total cost of the pilot project. The output from the pilot was sufficient for the executive to give the go-ahead to a full production, monthly customer profitability measurement, and reporting system using process-level costs.

IMPLEMENTATION PROJECT

Purpose

The objective of the implementation project was to deliver monthly, actual, process-based information that could be summarized into product and customer profitability. The information is used by:

- sales management to improve the profitability of trading relationships with customers,
- marketing management to determine product pricing, and

- general management to improve overall profitability using an Activity-Based Management approach to monitor cost and performance.

The full implementation enabled the reevaluation of outputs required from the system and the approach to be used. Due to the seasonal nature of the business, Willards decided to calculate customer profitability on a monthly basis. Willards determined that rather than developing a model at an activity level of detail, it would be quicker, more certain of success, and cheaper to develop the ABC model at the process level. A process could be broken down to the activity level of detail later, when the need arose. The spreadsheet software used for the pilot was not suitable for an ongoing production system. Oros software was purchased to handle the ABC calculations.

To achieve product, brand, market, and customer profitability, a two-stage approach was required:

1. Use the ABC system to cost the products (including marketing cost) up to the end of the Marketing process in the value chain, thus producing a cost per case of each product and pack size variant. In addition, use ABC to provide the outbound logistics and sales and distribution costs to get a case of product to the customer's outlet and to collect payment for the sale.

2. Use a data warehouse to combine the rates per unit of output measure for each process from the ABC system. Use the distribution and sales data from the operating systems to calculate the required profitability statements based upon actual customer sales for the period. A custom database was designed and built to handle the calculations needed to produce customer profitability data. PowerPlay and Excel were used to display the information and provide "slice and dice" and "what-if" capabilities. (The top-level information model design for the data warehouse is shown in Exhibit 15.1.)

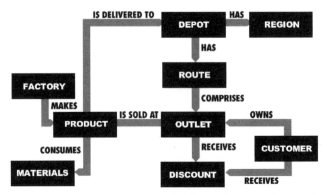

Exhibit 15.1 Willards Information Model (Reprinted with permission from National Brands Limited.)

CASE 1 163

Implementation Steps
The steps in the full project implementation were as follows:

1. Identify the source of all the data required for the ABC/warehouse systems. Develop download formats and procedures to dump data monthly into a designated directory on the file server. A reliable, flexible, and automated tool set is required to handle the collection of data, the calculations, and the export of the data to the data warehouse.
2. Design the Resource-Activity-Cost Object structure in Oros ABCPlus to calculate product and customer costs. Develop the import file structures to be able to import financial and nonfinancial data.
3. Develop a data model and build a data warehouse to process the sales and ABC data. Two key features of the warehouse are (1) the ability to reverse out erroneous data and reimport the correct data and (2) the ability to give each data set a name and to mix and match the data sets across the value chain. This gives the capability, for example, to assess the impact of a price increase for potatoes on June actual sales.
4. Design custom reports for users based upon their requirements using an EIS tool or Excel. This step was crucial for gaining buyer buy-in; giving the managers the opportunity to specify exactly what they wanted ensured their commitment.
5. Provide "goal seek" and "what-if . . ." capability for scenario planning. This gave the user the ability to flex the profitability calculations based upon changes in volumes, costs, and pricing, e.g. What if we decreased the price by 10 percent and got a 5 percent increase in volume across the board? How would this affect the brand profitability?

Outputs
In a typical profitability statement from the data warehouse, the source of the data is shown for each element. The EIS has a drill-down capability that provides the details behind each element on the statement. (See Exhibit 15.2.) The building block for this statement is the cost of manufacturing each product and the cost of getting it to the customer's outlet. As such, the profitability can be viewed along any of the dimensions shown in Exhibit 15.3.

Implementation Issues
Most times that people say they have implemented ABC, they are referring to a snapshot analysis. While a snapshot ABC project has its challenges, these are minor in comparison to the effort required to install a monthly running ABC program. A key success factor is ensuring the company has at least one person dedicated to the project—ideally the person who will take ownership of the system when the consultants leave. The speed with which the implementation of a monthly ABC system can be achieved is dependent on three issues: easy availability of data, quality of data, and the extent to which history is required.

A system has to be established to obtain the required data (normally nonfinancial)

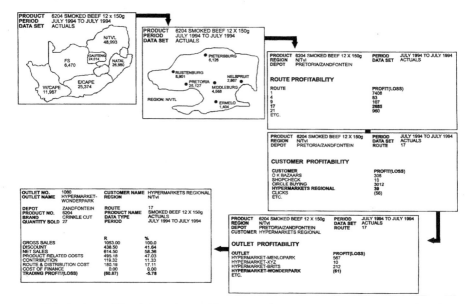

Exhibit 15.2 Willards Foods "Drill Down" (Reprinted with permission from National Brands Limited.)

quickly and easily. While it may be feasible to gather and input the data manually for a short while, ABC is only sustainable if this aspect is automated. This normally takes the form of modifying month-end runs of operating systems to download the required data into the ABC computer.

A major issue may be the inconsistencies of data across systems or data that has not been maintained properly in one or more systems. Although these errors create frustration, they are normally quite easy to sort out, particularly as IT in many companies move to a data warehouse concept for their MIS.

The extent to which history is required is the most insidious problem. Although it appears easy to provide six or more months of history for trend purposes, the effort required to gather and reconcile six or more months of historical data is such that it takes months, if ever, to catch up.

[My recommendation is to pick a starting month that is two months in the future, then use dummy runs to iron out any problems. If users are unable to complete the first month within the month, abandon it and use the next month as the start. Complaints about the lack of history are better than accusations of being out of date. Accurate and timely history soon builds.]

CONCLUSION

Many ABC implementations are designed simply to determine more accurate product costs. For this purpose snapshots are adequate; however, this limited use of ABC seldom

CASE 1

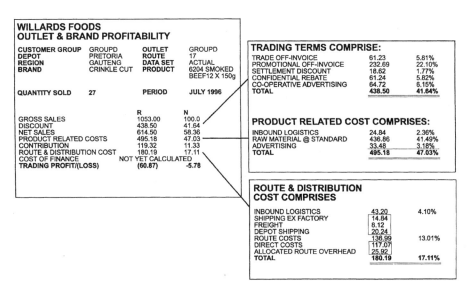

Exhibit 15.3 Outlet Profitability Statement to Supporting Detail (Reprinted with permission from National Brands Limited.)

provides value for money. Structuring the ABC project on a process basis and taking it to customer profitability enables decisions to be made that will pay back the project cost many times over each year.

Taking a process view of the activities provides strategic insight that is translated into bottom-line benefits. Willards Foods already has used the pilot's process and sample customer profitability information to restructure sales routes, close depots, and modify customer discount structures.

The effort to build a customer profitability database is not warranted if the information provided to users comes from an out-of-date snapshot of the business. The data and the database structures have to be kept up to date to reflect current trading relationships if the customer profitability information is to provide value for money.

The value of information is not in the reporting, but in the decisions that follow. ABC customer profitability information backed by a process view of both the costs and performance of the business provides the greatest value for money of all accounting information. It enables informed decisions on what to do, with what product, for which customer, and it provides insight for the holistic management of cost and performance.

CASE 2:
Providence Portland Medical Center Gets a $1.6m Shot in the Arm

By Don Miller, Providence Medical Center, Randall Benson Consulting, Rick Sahli, Providence Health System

HOW A NEW FAX MACHINE SAVES $500,000 A YEAR

When an in-house improvement team studied the medications process at a Portland hospital, they found a way to dramatically reduce the problem of illegible orders the pharmacy received each day. The team found imaging-enhancing fax machines (scan faxes) that could fax the faint NCR copy of a doctor's handwritten orders with clarity equal to the original. While their tests proved to be technically effective, the price tag was about $5,000 per fax unit, costing a total of over $100,000. This was more than ten times the cost of the fax units already in place.

The team members knew well that the hospital's Quality Council, under increasing pressures to reduce health care costs while maintaining and improving quality, would be wary of any recommendation that was not "low-cost or no-cost." So the team moved on to other improvement ideas, postponing a formal recommendation on the scan faxes until cost data was available. Eventually, the team was frustrated in its attempt to identify low-cost/no-cost improvements that would significantly improve the medications process (ordering, dispensing and administering medications on nursing units).

Aware of the team's frustration, Regional Director of Quality Faye Gilbarg recalled a presentation she had recently seen on activity-based costing and its potential in health care. She introduced the team members to Randall Benson, a consultant who was experienced in both ABC and process improvement. Benson helped them build and use an ABC cost model of the medications process. This model helped the team reach its improvement goals through some innovative cost analysis.

By tracing costs to the medication-related activities, they discovered that the scan fax machines would save over $500,000 per year in ordering, tracking and follow-up costs. ABC revealed that the scan fax machines, initially seen as too expensive, had a payback of only two months! Using ABC to search for more improvement prospects, the team identified over $1.5 million dollars in annual savings.

This is the story of that project. It tells how a process improvement team at Providence used ABC to gain new insights into the medications process and to exceed even their stretch goals for improvement.

INTRODUCTION

When Providence Portland Medical Center (PPMC), ranked one of the 100 best hospitals in the United States, recently formed a process improvement team, it knew it was tackling one of the most expensive and problem-prone areas in any acute care organization. That problem area lies in the ordering, distributing, and administering of medications to patients on nursing units.

The improvement team was made up of employees from nursing units, the pharmacy, and central supply. It was later joined by an ABC consultant. When the team decided to develop some useful ABC data, in addition to traditional process data such as turnaround time and quality indicators, they began to see more clearly where the hospital could save money and raise morale at the same time. While hopes were high that ABC would give some insight into the true cost of medication-related activities, no one could have guessed how much ABC would transform the work of the team.

Because the team could look at activity-based costs, it could see that some process improvements were dramatically more valuable than others. After discarding their original low-cost/no-cost rule of thumb, the team focused on high-cost activities and proposed improvements that would have been considered outrageous if not supported by the ABC.

Without the insights gathered from ABC data, the team could not have proposed that Providence should pull functioning fax machines from every nursing unit and pharmacy location, only to replace them with new models that cost nearly ten times more. Yet process innovations like these will allow Providence to save over $1.5 million per year on medication-related activities.

GENERAL BACKGROUND

Providence Portland Medical Center is a 483-bed, private, nonprofit community teaching hospital in Portland, Oregon. Health care delivery in Portland is highly competitive, with a large managed-care component. Health care providers have been asked by payers to reduce their costs dramatically. This has put pressure on all providers to find innovative ways of controlling costs, while still meeting customer expectations for high-quality care and service.

At Providence, these pressures have resulted in a drop in the historically high rate of employee satisfaction. One big contributing factor to this dissatisfaction was that employees were frustrated with the process of ordering and receiving equipment and supplies, including drugs and IV solutions. As a response, the hospital formed several quality improvement teams, including one to address this problem area.

THE PROJECT

Before hiring an ABC consultant, the team had used a mostly traditional process-improvement methodology. They mapped the process flow and surveyed employees, identifying three project goals:

1. reduce turnaround time from ordering to receipt of the medication/IV on the nursing unit;
2. improve communications about essential drug information (dose, route, etc.); and,
3. increase employee satisfaction with the process.

In its mission statement, the team chose not to state how much change they expected in these process goals, since no performance data existed. But team members could guess that, based on the amount of employee frustration and dissatisfaction with the process, the change would need to be significant.

Following traditional process improvement methods, the team's baseline measurement showed the magnitude of the problem.

- Only 28 percent of routine medications arrived on the nursing units within the expected one-hour time frame.
- A full 20 percent of orders were missing essential information.

When surveyed, employees were "very dissatisfied" with these process characteristics. The team knew they would be required to submit a cost/benefit analysis with any recommendations, despite the fact that traditional improvement methodology had never found the actual cost of the problem.

MODEL DEVELOPMENT

Because the team was already in full operation before they made the decision to use activity-based costing, time was short. To collect data, the team needed to design an ABC model that would help in these ways:

- identify medication-related activities that would be likely to reduce costs through process improvement,
- place financial value on their process-improvement concepts,
- figure out the impact on a cost-per-patient day,
- inform staff of the expected benefits of these process improvements, and
- make sure that resources were redeployed as planned.

To achieve these ABC modeling objectives in time, the team had to collect the resource, activity, and driver data in just a few weeks. With the help of their consultant, team members had to make some tough decisions about the breadth and precision of the data they would collect.

The team decided that the ABC model would cover 13 inpatient nursing units, representing about 80 percent of the total care units. It would also cover all pharmacy operations, central-services messengers, and nurses who set up IV fluid lines for patients.

Exhibit 15.4 shows the structure of the ABC model that the team built using Oros ABCPlus.

RESOURCES

The team obtained resource costs for these operations from 1996 operating statements. The statements were scanned into spreadsheets and then entered into Oros ABCPlus. While the operating statements did not show depreciation, they were, fortunately, free

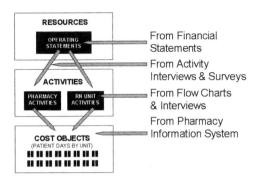

Exhibit 15.4 Medications Model Structure

of allocated overhead costs and were available for all departments. The statements also included the costs that were likely to be affected by process improvements, including labor, outside services, and supplies.

ACTIVITIES

The team developed the medication-related activity list from process flow charts they had completed earlier in the project. They supplemented those flow charts with interviews of staff members in the pharmacy and nursing units. The activity list covered only medication-related activities; all other activities were lumped into a category called "other non-medication."

The team also created attributes to identify activities that were not value-added or were particularly subject to errors, delays and rework. Nonfinancial information about the activities was just as important to the team as financial data. Finally, the team gathered percent value-added data and percent problematic attributes.

ACTIVITY DRIVERS

Activity drivers for staff-related costs were gathered through in-person interviews and hand-out questionnaires. The team used the factor of "percentage of staff time for each activity" to drive resource cost to activities. To save time, team members chose to collect driver data by staff type rather than by individual employee. In this way, they could directly associate the results with line items on departmental operating statements, avoiding the task of collecting and analyzing detailed payroll data.

Team members gathered resource driver data for nonemployee resources by interviewing front-line managers on nursing units, in central supply and in the pharmacy. Fortunately, most medication-related supplies are easy to associate with specific medication activities. For example, disposable syringes are always used during the "administer medusa" activity.

COST OBJECTS

The team struggled to define appropriate cost objects, mainly because it was unclear how cost-object data would be helpful for process improvement. The consultant asked the team members to think about what it was that the nursing units "sold," or produced. They decided that the nursing units produced patient-days of care. Supporting units, such as pharmacy and central supply, helped the nursing units create those patient-days of care. Each nursing unit, organized by diagnosis, produced a different variety of patient care. As a result, they arrived at medication related costs-per-patient day for each nursing unit as the cost objects for their study.

RESOURCE DRIVERS

The team then used pharmacy data to trace activity costs to cost objects. Most medication-related activities could be traced using either "number of medication orders filled" or "number of doses prepared." Both orders and doses were broken down into oral and intravenous categories and by shift (days, evenings, and nights).

Never before had Providence traced ancillary (in this case, pharmacy) activity costs to nursing units.

FINDINGS

Activity Costs

Overall, medication-related activity made up 43 percent of the nursing unit's total operating costs. These costs had never been spelled out so clearly. It was now clear that medication-related ABCs were by far the largest costs on the nursing units, and they knew that nursing unit costs were the largest single cost in the hospital.

Clarifying the largest costs like this galvanized the team members. At that point they could see they needed to go beyond their unofficial ethic: to look only for improvements that were low cost or no cost.

Exhibit 15.5 shows the cumulative medication-related activity costs (which excludes the cost of the medications themselves) for all the participating departments. The team, which had been focusing largely on pharmacy activities, was surprised to see that, according to the ABC data, most of the cost leverage was not in the pharmacy where drug orders are filled, but on the nursing units where the medications are ordered and administered. The team decided to shift its emphasis from saving pharmacy time to saving nursing time.

In particular, the team focused its attention on order-placing and follow-up activities. Not only were these drivers a large part of the total medication-related costs, but they also greatly increased the non-value-added cost.

Cost Object Costs

In the medication process, the customer is served on the nursing unit, not at the pharmacy or in central supply. Thus the team focused on the nursing unit as the point where the care product is delivered. Each nursing unit is organized by patient diagnosis and/or

CASE 2

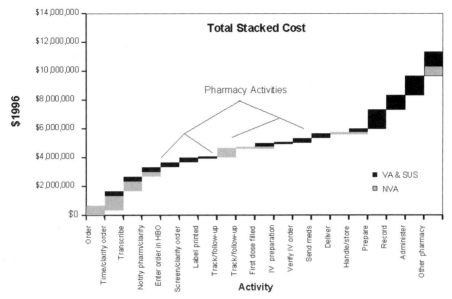

Exhibit 15.5 Cumulative Activity Costs

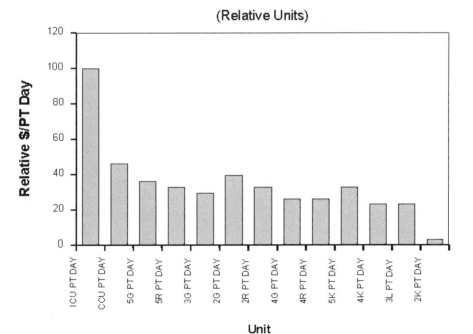

Exhibit 15.6 Outcome (Cost-Object) Costs

intensity of care. The team used "medication costs per patient day" as their cost objects, with one cost object for each nursing unit. Exhibit 15.6 shows the results of the ABC analysis.

Once again, the ABC analysis added critical insight into the medication process. The intensive care unit (ICU) had much higher medication costs per patient day than the other units, even those units that also cared for very sick patients. By working backward from the cost object module to the activity module of Oros ABC, team members could see that the ordering activities cost substantially more in the ICU.

When they went back to the ICU, expecting to discover a data collection error, they were told by front-line caregivers that the numbers made sense. These caregivers reminded the team members that the ICU is the key teaching unit in the hospital, with physician interns and first-year residents writing many of the medication orders for patients. The orders in ICU are complicated, and the order-makers spend more time writing orders, often including other clinicians in the process. As a result, the ordering activities take more staff time on the unit and more pharmacy time to alert ICU for possible problems, such as drug interactions, allergies, or preferred alternate drugs.

Process Improvement

The team members used the ABC information to find where process improvements would drive down cost and turnaround time while adding to employee satisfaction. If the cost improvement potential was high, then they could discard the low-cost, no-cost rule of thumb. Now they could see where to spend money, if necessary, to improve high-cost activities.

For example, the team members knew that the main reason for notifying the pharmacy of new orders and tracking and following up on the orders was illegibility of the faxed physician's order. They could not send the order via e-mail or database update, because the pharmacist is required to use a copy of the physician's handwritten order, not a computer summary of the order. Earlier in the project, they had looked at image-enhancing fax machines, or scan faxes. It worked better than anyone had hoped, but the cost of about $5,000 each was deemed too high, given that the ordinary fax machines already in place only cost about $500 each.

But trials showed that the scan fax virtually eliminated unreadable orders and reduced follow-up phone calls by over 90 percent. Exhibit 15.7, with a typical nursing unit and all units taken together, shows how economical the scan fax would turn out to be in the long run. Given the huge cost-savings potential, the annual cost of the scan fax was easily justified. ABC analysis revealed that the improvement idea was a huge improvement opportunity that can save the hospital over $500,000 per year.

The team evaluated dozens of other process improvement concepts. Eight of these were estimated to save between $250,000 and $700,000 per year in activity costs. After analyzing the concepts separately, team members combined these ideas into five improvement scenarios. Every scenario included the scan fax concept. They evaluated each on the basis of activity-based cost, turnaround time, and employee satisfaction. Narrowing the list, they eventually reached a preferred scenario which they recommended to the Quality Council.

CASE 2

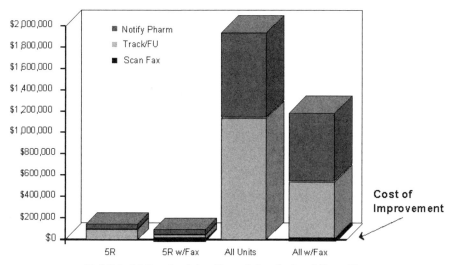

Exhibit 15.7 Activity-Based Analysis of Scan Fax

Once ABC analysis had been completed, it was clear to team members that the cost savings more than offset the cost of the scan/fax machines. Because the trial had been so successful, and because the financial cost benefits, as demonstrated by ABC methodology, were so compelling, the team went forward with this recommendation even before making its final report to the Quality Council. The recommendation was approved and almost immediately put into effect. When employees found that the new scan fax machines reduced time and aggravation on the nursing unit, many of them said that it was one of the best ideas to have come out of a quality improvement team.

RESULTS

Using brainstorming techniques, the team generated a list of possible process changes to reach their three original process goals: improved communications about drugs and IV solutions, timely medication delivery, and increased employee satisfaction. The brainstorm list was prioritized, based on its ease of implementation, its potential effect on the problem, and its feasibility.

The first recommendation to come from this list, the scan fax machine, reduced time and frustration, increased employee satisfaction, and improved both communications and timely medication delivery. The balance of the team's recommendations, made up of equipment, personnel, and remodeling, came to a total implementation cost of over $400,000 per year. However, the potential cost savings in time that would no longer have to be spent on some of the medication-related activities amounted to over $1.6 million. Thus the net savings in redeployable resources was over $1 million. This amount translates to additional time that nurses and pharmacists can spend on direct patient care.

NEXT STEPS

The team plans to implement the additional recommendations in several phases through the second quarter of 1998. The timing of the plan is based on sequencing of training, remodeling schedules, and the purchase of equipment.

Over the years, the Quality Council at Providence Portland has been concerned that savings identified by improvement teams often do not materialize. Even after improvements are put into effect, savings cannot always be confirmed. To allay these concerns, the medications-process improvement team has committed itself to updating the resource drivers by surveying employees after the improvements are in place. Running the ABC model again with updated drivers should confirm that nursing-unit resources were redeployed from medication-related to other care activities.

LESSONS AND CONCLUSIONS REGARDING ABC/ABM

The improvement team, the Quality Council, and the administrative sponsors at Providence have come up with several lessons and conclusions from this first attempt to apply activity-based costing:

- The improvement-team members could easily handle ABC data. They could immediately use the cost data in the same way as any other process or quality data. The team found it much easier to interpret ABC data than traditional financial statements; there was virtually no learning curve.
- ABC helped the team understand the causes of costs (that is, the activities), not just the magnitude of the costs.
- ABC allowed the team members to look at costs across the entire process, freeing them from the trade-off thinking implied by departmental budgets.
- The team had planned to use ABC for cost-benefit analysis but quickly extended its application, using ABC to identify some previously-unseen improvement opportunities.
- Informed about activity costs, the team dropped its low-cost, no-cost rule of thumb to think big about high-leverage solutions that could produce much bigger improvements.
- The team can verify resource redeployment, using updated ABC drivers. This will be a major innovation, especially since financial improvements from earlier projects have been hard to confirm.
- Charged with process improvement, the team was much more interested in activity costs than cost-object costs. Yet even a cursory cost-object analysis highlighted some sizable process improvement opportunities in the ICU.

CONCLUSION

The use of ABC at Providence Portland transformed process improvement for the medication team. Armed with the cost dimension of their process, the team members could, for the first time at Providence, use cost in the same way they used operational and quality data to find and gauge improvement opportunities. The result was most likely an order-of-magnitude increase in the value of the improvements they recommended, moving from tens of thousands to over $1.6 million in improvement recommendations.

CASE 3:
US Airways Takes Off with ABC

By Joe Donnelly, Arthur Andersen & Co. LLP, and Dave Buchanan, US Airways

IMPLEMENTATION LANDS $4.3M IN PROCESS IMPROVEMENT SAVINGS

US Airways is one of the world's leading air carriers and one of its largest. It is also one of the largest U.S. airlines and holds the dominant market position in the eastern U.S. With about 42,000 employees, the company operates more than 2,000 mainline flights daily, serving more than 100 airports in the United States, Canada, Mexico, the Caribbean and Gulf of Mexico, Germany, Italy, France, and Spain.

The US Airways system also includes the US Airways Shuttle and US Airways Express, and the system as a whole operates more than 4,500 daily departures to more than 200 airports.

The US Airways Powerplant Department, or aircraft engine shop, at the Pittsburgh International Airport, maintains and overhauls 350 jet engines powering 152 of US Airways' 380 aircraft. When an engine arrives at the US Airways Powerplant Department for maintenance, it can have as many as 11 separate modules, which can be repaired or overhauled. In-house repair and overhaul capabilities focus on the seven model types of the Pratt & Whitney JT8 engine family. The engine shop operates 24 hours a day, five days a week.

There are about 500 employees in the engine shop. The workforce is divided into 24 separate, self-directed work teams. The majority of the powerplant employees are represented by the International Association of Machinists (IAM). US Airways and the IAM implemented a High Performance Work Organization initiative whereby they establish a collaborative work environment with common goals. Each work team develops its own charter and selects a team leader. Team members actively manage their work areas and identify opportunities for improvement.

UNDERSTANDING THE BUSINESS NEEDS

US Airways needed detailed cost information with particular focus on engine overhaul costs. The lack of detailed operational and financial information did not allow management to fully understand the costs associated with producing or overhauling an engine, and, to a lesser extent, the costs of each of the modules making up an engine. This led management to ask other key questions which would enable them to determine the business solution needed to provide this information. Some of the key questions included:

- What are US Airways' critical business issues;
- What industry trends are driving these issues;
- What is the strategy to meet the needs of the changing business environment;

- What operational and financial information would provide decision support to meet these business needs;
- What level of detail is needed to manage; and,
- What frequency of reporting is needed?

The answer to each of these questions would provide insight to the action needed to deliver the necessary information. US Airways determined that more detailed and insightful information was needed to manage the business, drive improvement, manage costs, and support third-party pricing. The business solution would have to support each of these needs.

BUSINESS SOLUTION

Activity-based cost management (ABCM) was the business solution chosen for many reasons. ABCM could not only help determine the true cost of engine maintenance, but could also provide operational and financial information to be used by the self-directed work teams to identify opportunities for improvement. ABCM would also be the key enabler for US Airways to better understand the following:

- Repair in stock vs. buy new—inventory decision making;
- True cost of operations to support third party pricing;
- Operational metrics for improvement and bench-marking; and,
- Impact of improvements made by self-directed work teams.

An ABC team was formed at US Airways to implement ABCM in the powerplant department. The team was composed of two full-time US Airways employees from the engine shop, a US Airways financial staff member, and two Arthur Andersen members.

PROJECT APPROACH

The project was divided into four main phases. The phases consisted of introduction to ABC, data collection and information gathering, model building, and data analysis and reporting.

Introduction to ABC Phase

To ensure a common starting point, several meetings were held with all employees to explain the ABC project, why it was being undertaken, and the insights it would provide. These initial meetings served to achieve employee buy-in and support. Subsequent meetings were held with team leaders and representative members of each of the 24 work teams who were most familiar with the engine shop processes and who could describe the objectives and the benefits of the initiative.

Data Collection and Information Gathering

One of the initial steps in the data collection phase was to understand the level of cost object detail to be costed. By understanding the level of detail to be costed, the team could gain insight into the types of activities and the level of detail to be collected and measured. Once the products to be costed were understood, the team began to gather specific cost pool information. The team determined it was necessary to collect and include the cost of rent, utilities, and benefits, which were paid by the corporate headquarters, in order to determine the true cost of the powerplant department. Costs were also collected for other areas known as the periphery groups which include Warranty, Stores, Training, and Purchasing. These are separate groups within US Airways who operate their own budgets and perform services not only for the powerplant department, but other areas as well. These costs were also important to include in the analysis in order to understand the true cost of the powerplant department.

Next, activities were collected from each of the work teams, including core and non-core activities, such as tear-down, welding, waiting for tooling, and rework. The diversity of the workforce and layout of work areas into teams made collecting the necessary operational and financial information challenging for the team. The final activity dictionary included 410 activities across the engine shop, including 47 non–value-added activities. Representatives from each of the 24 self-directed work teams were then interviewed to determine how their effort was distributed across their team specific activities. The activities were captured on team-specific spreadsheets and populated by the representatives based on percentage of effort. In addition, the spreadsheet design captured any cost diversity between the different engine types, which illustrated if certain engine model types required more effort for a certain activity. To assist the employees in completing the effort spreadsheets, a time conversion table was developed to convert the actual time spent on activities into percentages. The periphery groups that performed services for the powerplant department were also included in this process in order to determine the number of employees dedicated to powerplant tasks and the associated effort spent on those tasks. The team then attributed each of the activities as either core, support, contractual, or irrelevant activities to understand how the powerplant was investing its time and effort.

Across all four phases of the project, project management controls were used to assist in defining the scope of the tasks and progress being made in completing them. These controls were essential considering activity information was needed for 500 people working three shifts, with 12 separate classifications of employees such as mechanics, welders, machinists, and inspectors—most with different labor rates. Obtaining the final cost and headcount per process was complicated as teams of people were dedicated to many different processes. A headcount control sheet allowed the cost management team to identify the true headcount associated with each of the processes.

Another project control included a progress chart. This chart detailed each task the team needed to be accomplished for each of the 24 work teams. As the tasks were completed, the color-coded chart was updated. Using visual controls, any member could quickly see what remained to be accomplished as well as the scope of the remaining tasks.

Model Building Using Oros

Building the model using the Enterprise Pack software, including Oros ABCPlus, EIS, and the Links Engine, the US Airways ABC team constructed its entire model with import files. The first module to be constructed was the cost object module based on the level of detail determined in the data collection phase. The cost object hierarchy would accommodate the variability in levels of service for each engine type. Each engine has different maintenance needs or service levels; the model allowed US Airways to determine the true costs for each type of engine by each service level. The activity module was built next. The activity dictionary, developed earlier in the data collection phase, was easily imported into the activity module from a spreadsheet format. Finally, the resource module was built. The resource hierarchy was also imported from a spreadsheet format based on the level of detail determined earlier in the project. Driver data was imported after each module structure or hierarchy was in place in the model.

The cost management team built intuitive abbreviated names and initials into its model reference numbers to help them diagnose any problems they might have during model building. In doing so, the team could immediately determine where specific account dollar amounts were derived. The use of these intuitive reference numbers proved to be invaluable during error and warning diagnosis by enabling the team to quickly pinpoint the problem areas. Once the model was running, it provided true cost information previously unavailable within the powerplant department.

Data Analysis and Reporting

After the model was built and the ABC data was available, the data needed to be analyzed and reported. The team first conducted reasonableness checks to ensure the model was assigning costs appropriately. Next, the team had to assimilate the data into a format that could be reported and easily understood by the engine shop employees. Report books were created for each of the 24 self-directed work teams with their individual results as well as a view of the entire engine shop. These reports were well received by the teams and validated their contributions during the data collection phases. Team leaders as well as all team members now had access to operational and financial information that would enable them to drive improvement, measure results, and cultivate process ownership.

PROJECT RESULTS

The ABC model output provided US Airways with operational and financial data to support strategic and operational decision making. The ABC information identified process improvement savings opportunities for US Airways totaling $4.3 million per year and 63 full-time equivalents. The model output provided numerous operational and financial metrics which were not previously available. The self-directed work teams could now see the activities' cost by labor classification and by shift. In addition, a rank order analysis for the activity costs was conducted. This view provided insight into the most expensive non-value-added activities occurring in each work team. This information was immediately available to begin analysis for process improvement opportunities. For example, one work team found they spent about 80 percent of their effort or $85,000 per year reinspecting due to the use of an older piece of inspection equipment. The ABC information justified the purchase of the new piece of inspection equipment, the cost of which was less than the $85,000 spent on reinspection each year.

The team, now equipped with operational and financial information, needed to capitalize on the identified opportunities in response to the business issues set forth at the beginning of the project. The ABC model output could now be used to support the business through:

- Support of repair vs. buy inventory decisions;
- Understanding the true cost of operations to support pricing;
- Supporting benchmarking; and,
- Providing a baseline for measuring improvements by the self-directed work teams.

Although irrelevant activities were identified, root cause analysis would have to be undertaken by the team to identify the causes for the effort. Upon identifying the true cause, the company and teams will be able to eliminate or reduce the impact of non–value-added activities. The ABC team now publishes a unit cost guide three times a year. This guide details per unit costs for each of the seven engine types and their modules.

NEXT STEPS

Powerplant management made the decision to update the cost model on a trimester basis. Therefore, every four months, headcount for each of the teams and processes will be updated in the model. Plans also call for annual activity and effort updates via the effort grids, which will provide progress reports on process improvements made by the self-directed work teams in the engine shop. When the engine shop undertook this ABC project, there were a number of shortcomings in the ability to provide necessary operational and financial data for the cost model. For example, while the company knew how many engines it produced, there was not accurate information on how many modules were produced. As a result, the team set up procedures to capture accurate production by engine and by module. The ABC team now publishes a monthly engine and module production summary for the powerplant department. The team also found weakness in tracking for overtime costs. Many of these costs were grouped together, which did not allow a view of what team or discipline actually incurred the costs. The cost management team brought greater detail to these costs by using an overtime report, also published monthly.

The need for accurate and timely data to support the cost model will drive the team to continually validate their data frequency, source, format, and capture methods. Communication of results is an ongoing effort. The team will need to communicate ABC results to the people who can and will implement change.

CASE 4:
Grupo Casa Autrey's CFO Drives Profitability Using ABC

An interview with Rubén G. Camiro, CFO of Grupo Casa Autrey on how ABC has made a difference.

By Jorge Medina More E.

With a clear idea of the business, use of focused software, and a smart corporate strategy, you can model the functionality of the enterprise. This is the case of Grupo Casa Autrey, which since 1996 has invested in ABC methodology to solve their costing problems and create a fast, reliable, and current information tool for the operational and managerial decision-making process.

As one of Mexico's leading wholesale distributors and with 104 years of experience in the domestic market, Grupo Casa Autrey delivers and markets pharmaceutical and consumer products, including home, health and beauty products, nonperishable foods, publications, and office supplies. Currently its product catalog contains over 12,000 items and its operations are handled by 7,000 employees in 24 distribution centers through out Mexico. Each month Grupo Casa Autrey distributes more than 40 million products to its customers. 1997 sales revenues were 8,746 million pesos (more than 1 billion U.S. dollars), a 13.8 percent increase above previous year. "This is the result," says Rubén G. Camiro, "of having a clear corporate mission and a vision we define as being the preferred supplier to our customers, and the preferred customers to our suppliers."

Grupo Casa Autrey's reputation is based on their commitment to the highest quality service. It is this commitment that has made them one of the most efficient and reliable distributors in Latin America.

In the interview that follows, Rubén G. Camiro CFO at Grupo Casa Autrey explains more about the use of ABC at Grupo Casa Autrey. Camiro, who promotes and coordinates the ABC project, has had a brilliant career for 14 years with the organization and has held his current position for two years.

Why Did Grupo Casa Autrey Choose the ABC Methodology?

The objective for implementing ABC was to better understand our company from every angle, because over time we have become a very diversified organization. When we were a one-channel enterprise and only dealt with pharmaceutical products, it was relatively easy to measure our performance. But handling various types of customers, with multiple products and services, makes the measurement criteria and decision making more complex. We had to have an accurate way to measure our activities and processes. The bottom line was to have a better understanding of our profitability in the different business units.

What Changes Did You Have to Make to Adopt the ABC Methodology?

Grupo Casa Autrey has always had a clear philosophy about information management, in which ABC fits quite naturally. The most meaningful change was in accounting. We adopted a multidimensional accounting process and redefined the cost centers in accordance with the ABC requirements. This allowed us to easily implement activity-based costing.

What Benefits Have You Seen Since Introducing ABC?

Great benefits were seen beginning from installation. While reviewing all of the company processes and evaluating how standardized and documented they were, the details we were looking for started to appear. There were some areas that weren't totally automated in the information gathering process, and now they are. Through these actions we have achieved a great deal in savings and efficiency. Benefits have resulted in greater productivity in the warehouse, delivery routes, and every area of the company. For example, in the past some believed that certain customers were not profitable. We have now identified the profitable ones and we know how we can convert the nonprofitable ones to into profitable ones. We've identified the causes and designed ways to modify behaviors to result in a successful business relationship.

To What Degree Has a Better Knowledge of the Organization Been Achieved?

The group of directors who participated in the ABC model development have a similar knowledge level of the enterprise. We learned that in some departments, responsible personnel did not understand more than 1 or 2 percent of their processes. Having 100 percent knowledge has been very positive.

When we compared our first study using ABC with our final ABC model, our perception of the company changed radically. We moved from seeing ourselves as a highly complex organization, to one with simpler and more linear processes. In the first study we identified more than 490 activities; currently there are only 97. An interesting benefit is that we have consensus regarding what is a procedure and what is a method. Though we still have a way to go, we can validate that we are on the right track.

How Does an External Consultant Help on This Type of Project?

ABC is still an art in some respects. It is a matter of methodology and order. It is very easy to get lost in the details if you do it alone. Not because it can't be done, but it is helpful to have a critical and focused outside opinion. In something relatively new, an expert opinion is invaluable. The consultants of Arthur Andersen contributed a great deal, by helping us understand how deep to delve into the issues.

What Obstacles Did You Encounter during Implementation?

In the technical implementation there were obstacles to overcome, interfaces that needed to be generated, and other technical problems to be solved. These problems were anticipated, but they were few and workable. The main difficulty was that people were not accustomed to coding their activities in detail. It was an attitude problem of breaking obsolete paradigms. It was interesting to see we all did not have the same vision of the enterprise. Creating our model was an exercise in standardization. On my part, it has also been a matter of convincing everyone to take advantage of this technology. We no longer accept past mistakes. Where there are still minor errors, they are easily identified and corrected due to what we have learned. With the adoption of ABC, the benefits have outweighed the implementation difficulties, and things now flow much more smoothly and there is consistency through the enterprise.

What Comparative Advantages Have You Experienced Since Implementing ABC?

We have always excelled in service, quality products, and capacity to live up to the market's demands. We cannot afford to be anything but an efficient provider in a market that is consolidating, as the pharmaceutical market is. We are a company that provides service to its customers. The market demands it and keeps recognizing what Grupo Casa Autrey does.

Our business model continues to be different than that of our competitors. The fact that we know ourselves better gives us an advantage. Now that we are in the final stages of the model design, and cost and driver analysis, we can set up goals and establish actions to reward our customers in accordance to their performance. This gives us a new advantage which is allowing us to be more focused.

Would You Describe Your Model as Strategic or Operational?

Conceptually, the model is strategic. It begins with business strategy, making the big decisions and then defining policies. Once the model is established, it takes on operational characteristics because of the richness of the information it handles. The database contains every transaction, which provides a range of drivers. We can analyze activities in detail and tune up our knowledge of the processes. The model is simple in scope but deep in reach.

How Do You Transfer the Benefits of This Methodology to the Organization?

At every level of the company we experience the benefits of using ABC and examining current and reliable information. How the overhead is assigned to each business unit always causes debate. This directly impacts each area and motivates people to take the initiative to look for that information. Departments can verify why they have been charged with a particular activity, and drives a desire to better understand the model. The same thing happens with monthly reports, where the departments see the profitability information.

What Additional Uses Does Grupo Casa Autrey Have in Mind for ABC?

On the strategic side, the question of how to create new businesses or reorganize current ones recurs. One of the great challenges in the medium term is to determine the value of holding to our current structure and evaluating the installed capacity. The answer is not simple, but with ABC we have the necessary information to deal with this issue.

On the other hand, for the continuous internal improvement process, the current ABC model provides an economic map of the organization. Using the process view analysis, we can establish the necessary performance policies to produce improvement. The performance indicators based on our activities are dynamically linked to our balanced scorecard.

Do You Have Any Closing Comments?

It has been a long road, and along the way we have made some mistakes. The achievements, however, outweigh the mistakes and provide a solid platform on which to base important decisions.

A big accomplishment was restructuring our general ledger system, which can now be updated constantly in a very simple way. Details are automated and results are reliable and standardized. The richness of our ABC model is demonstrated by its ability to measure at the lowest possible level the activities and cost objects. Since making the decision to use this methodology, we have experienced a significant change in what to expect from ABC. Our initial objective was to cost clients and products. The current objective is to cost transportation routes, sales agents, and products, which have an obvious impact on the service to our customers. Acknowledging this has allowed us to design better corporate strategies.

In three or four months, this model will give us precise results and measurements, enable us to improve the personnel performance, and allow us to set new incentive policies. The best is yet to come.

NOTES

1. Reprinted with permission from ABC Technologies Inc. and Willard Foods. K. Phillips and Kevin Dilton-Hill, "Willards Foods: Managing Customer Profitability with ABC Information," *As Easy as ABC: ABC Technologies Newsletter* (Winter 1996).
2. Reprinted with permission from ABC Technologies Inc. Don Miller, Randall Benson, and Rick Sahli, "Providence Portland Medical Center Gets a $1.6 M. Shot in the Arm," *As Easy as ABC: ABC Technologies Newsletter* (Fall 1997).
3. Reprinted with permission from ABC Technologies Inc. J. Donnelly and Dave Bucanan, "US Airways Takes Off with ABC," *As Easy as ABC: ABC Technologies Newsletter* (Winter 1997).
4. Reprinted with permission from ABC Technologies Inc. and Grupo Casa Autrey. Jorge Medina More E., "Grupo Casa Autrey's CFO Drives Profitability Using ABC," *As Easy as ABC: ABC Technologies Newsletter* (Summer 1998).

APPENDIX A

CHECKLIST FOR SYSTEM SELECTION

This checklist can be used to understand and compare activity-based information systems from vendors. It is not meant to be an exhaustive list but is meant to be a starting point for discussions. It is also designed for users to establish their biases using weighted measures toward certain features within an ABIS. Some suggestions for use are:

- Weight from 1 to 5, 1 being low, 5 for high, the relative value of the features listed.
- Score from 1 to 5, 1 being low, 5 for high, scoring the feature provided by the vendor.
- Multiply the score by the weight to derive the resultant weighted score.
- In each category, such as infrastructure, products, and so on, compare vendor A to vendor B to isolate a priority between them.

Vendor Selection Criteria

INFRASTRUCTURE	Company A			Company B		
	weight	score	result	weight	score	result
Business Model						
Public/private company						
Profitability						
Consulting firm? or information sys company						
D & B rating						
do you work with my customers?						
price/performance						
web page						
knowledge company /software company?						
demonstration						
demo disk						
people power						
culture?						
GSA rated? (government endorsement)						
industry rated?						
product warranties						
internet?						
enterprise-wide rollouts experience						
visited company?						
% revenue service vs. software						
local R & D						
Alliances						
Universities/academic						
Implementation support from Big Five firms						
industry-specific consultants						
alignment with thought leaders						
Focus on ABC						
# support engineers						
% dedicated to R & D						
# of developers						
Years of experience in ABC						
Multilocation						
Management processes						
One-product company						
International User Group						
# of users attending						
# of global locations to attend						
Keynote speaker quality						

… APPENDIX A

Vendor Selection Criteria *(cont'd)*

INFRASTRUCTURE	Company A			Company B		
	weight	score	result	weight	score	result
# returning to user group yearly						
# of case studies vs. company presentors						
# relating to my industry						
Installed Base						
size						
Local installed base reference						
Global focus						
References by industry						
References total						
% repeat business						
Track record of implementations						

Product Focus

	Company A			Company B		
	weight	score	result	weight	score	result
Integrated to ERP						
Stand-alone						
Connectivity						
Integration						
SW quality tracking system						
Product enhancement system						
Operating systems						
32-bit architecture						
First release?						
Network version						
Multiuser						
Structure optimization						
Structure integrity						
Data integrity check						
Calculation times						
Access controls						
COM/DCOM architecture						
IS acceptance						
Model-to-model linkage						
Response times						
3-tier client-server						
Application C++						
VB interfacing						
HW requirements						
Open database						
open standards						

Product Focus *(cont'd)*

	Company A			Company B		
	weight	score	result	weight	score	result
structure						
size						
proprietary?						
User Interface						
CAM-I standard						
Open architecture						
standards based?						
warranty						
maintenance plan						
on-line help						
on-line consulting						
licensing scheme						
manageable						
scaleable						
maintainable						
security						
password						
levels						
In-model documentation						
General tutorials in sw						

Information Capture

INFORMATION CAPTURE FEATURES AND DEPLOYMENT

	Company A			Company B		
	weight	score	result	weight	score	result
ASCII data importing/exporting						
multiformating						
flexibility						
Links application to databases						
ODBC connectivity						
scripting						
access drivers						
query language						
SQL based						
Natural language based						
emote automation						
Connection to spreadsheets						
2 way?						
limits in sheetlength						
3rd-party linkages						
GUI						

APPENDIX A

Information Capture *(cont'd)*

	Company A			Company B		
	weight	score	result	weight	score	result
windows						
process views						
graphical						
updating capability						
ease of use						
ease of approach						
CAM-I standard terms						
custom views						
Electronic data capture						
connected to modeling engine						
web-based capture						

Modeling and Analysis

	Company A			Company B		
	weight	score	result	weight	score	result
Activity/resources/cost objects						
limits on amount						
attributing						
activity dictionary						
activity clustering						
Drivers						
numeric						
nonnumeric						
equational						
weighted						
negative						
Process modeling						
Structure						
flexibility						
update friendly						
Multidimensionality						
# of dimensions						
limits to dimensionality						
User-defineability						
GL interfacing						
Periods						
Built-in reports						
Auditing						
Bill of activities						
Bill of materials						
Revenue tables						
Modeling methods						
CAM-I						
Bill of Cost						

Modeling and Analysis *(cont'd)*

	Company A			Company B		
	weight	score	result	weight	score	result
Notes functions						
Revenue tables						
Profitability						
Named views of models						
Yield analysis						
Capacity analysis						
Depth of hierarchy in files structure						
Backward calculations						
Hierarchy and roll-ups						
Automatic saves						
ABB						
Target costing						
Linkage to strategy						
Performance measures						
Scripting/automation						
Linkage to KPIs						
Levels						
Configurability						
Named views						
Model recalibrations						
Interactive validation						
Rules checking						
Error checking						
Filtering						
Multilanguage						
Multicurrency						
Multimodel						
Multiuser						
Read/write						

Reporting and Data Navigation

	Company A			Company B		
	weight	score	result	weight	score	result
Built-in reports						
Graphing/charting						
Custom reporting						
Report mining						
Data navigation						
automatic drilldown						
OLAP programming						
slicing & dicing						
Print preview						

Reporting and Data Navigation *(cont'd)*

	Company A			Company B		
	weight	score	result	weight	score	result
Reports						
contribution						
multilevel contribution						
history						
Trending						
Benchmarking						
HTML interfaces						

Scenario and What-If Subsystems

	Company A			Company B		
	weight	score	result	weight	score	result
What-if reports						
Scenario building						
Run model in reverse						
Survey connectivity						
Survey mining						

Services

	Company A			Company B		
	weight	score	result	weight	score	result
Training						
on-site						
public						
Internet-distance learning						
materials						
trainers						
computer-based training						
Support						
technical talent						
escalation procedure						
quality metrics						
telephone support						
worldwide consistency						
on-line assistance						
# of support staff						
locations						
24-hr close rate						
# calls per week						
internet accessibility						
24-hr support line						

Services *(cont'd)*

	Company A			Company B		
	weight	score	result	weight	score	result
Updates						
Upgrades						
Fee						
Newsletters						
# case studies vs. propaganda						
quality						
industry participation						
Educational Materials						
books						
videos						
white papers						
Web page access						
Client Service						
resident experts						
expert modelers						
global consistency						
# models developed/yr						
total years of experience						
industry-specific knowledge						

APPENDIX B

INFORMATIONAL WEB SITES

www.abctech.com	Software company/reference library/books/videos
www.armstrong-laing.co.uk	Software company/consulting firm
www.arthurandersen.com	Arthur Andersen LLP
www.Baan.com	ERP vendor
www.cam-i.org	Consortium of Advanced Manufacturing-International
www.deloitte.com	Delloitte & Touche LLP
www.ey.com	Ernst & Young LLP
www.icms.net	Software company and consulting
www.kellogg.nwu.com	Kellogg Graduate School of Management
www.leadsoftware.com	Software company
www.people.hbs.edu/ckaplan	Professor Bob Kaplan biography
www.peoplesoft.com	ERP vendor
www.pwcglobal.com	PriceWaterhouse Coopers LLP
www.sap.com	ERP vendor
www.sapling.com	Software company
www.us.kpmg.com	KPMG Peat Marwick LLP
www.wiley.com	Books

SUGGESTED READINGS

BOOKS

Activity-based Management Process Improvement Group ECR Operating Committee, Ernst & Young, LLP. *Activity-based Management: Case Studies* (Washington, D.C.: Joint Industry Project on Efficient Consumer Response, 1997).

Balachandran, Bala. *Strategic Activity-Based Accounting* (New York: Prentice-Hall, 1994).

Biudion, Claude, and Glenn Hollowell. *Realizing the Object-Oriented Lifecycle* (Englewood Cliffs, NJ: Prentice-Hall PTR, 1996).

Cohen, Stanford, Doris Kelley, Daniel Ford, and Mary Galvin. *Utility Industry—Armed to the Teeth* (New York: Merrill Lynch, 1994).

Cokins, Gary. *Activity-Based Cost Management—Making It Work* (Chicago: Irwin, 1996).

Connelly, Robin McNeill, and Roland Mosimann. *The Multidimensional Manager: 24 Ways to Impact Your Bottom Line in 90 Days* (Canada: Cognos Inc., 1996).

Covey, Steven. *The Seven Habits of Highly Effective People* (New York: Simon & Schuster, 1990).

Dragoo, Bob. *Real-time Profit Management—Making Your Bottom Line a Sure Thing* (New York: John Wiley & Sons, 1995).

ERC Performance Measure Operating Committee. *Performance Measurement: Applying Value-Chain Analysis to the Grocery Industry* (Washington, D.C.: Joint Industry Project on Efficient Consumer Response, 1994).

Grove, Andrew. *High Output Management* (New York: Random House, 1983).

Hronec, Steven M. *Vital Signs* (New York: AMACOM, 1993).

Johnson, H. Thomas, and Robert S. Kaplan. *Relevance Lost: The Rise and Fall of Management Accounting* (Boston: Harvard Business School Press, 1987).

Kaplan, Robert S., and Robin Cooper. *Cost & Effect* (Boston: Harvard Business School Press, 1998).

Kaplan, Robert S., and David P. Norton. *The Balanced Scorecard* (Boston: Harvard Business School Press, 1996).

Marberley, Julie. *The PriceWaterhouse Guide to Activity-Based Costing for Financial Institutions* (Chicago: Irwin, 1992).

Miller, John. *Implementing Activity-Based Management in Daily Operations* (New York: John Wiley & Sons, 1996).

O'Guinn, Michael C. *The Complete Guide to Activity-based Costing* (Englewood Cliffs, NJ: Prentice-Hall, 1991).

O'Toole, James. *Leading Change* (San Francisco: Jossey-Bass, 1995).

Ostrenga, M. R., Tarrence R. Osan, Robert D. Mcilhartan, and Marcus D. Harwood. *Ernst & Young Guide to Total Cost Management* (New York: John Wiley & Sons, 1992).

Player, Steve, and David Keys. *Activity-Based Management—Arthur Andersen's Lessons from the ABM Battlefield* (New York: Mastermedia Ltd., 1995).

Ribler, Karen, and Deb Dixon. *Activity-based Management: A Primer for Foodservice Brokers* (Reston, VA: Association of Sales and Marketing Companies Foundation, 1996).

Senge, Peter M. *The Fifth Discipline* (New York: Doubleday Currency, 1990).

Stratten, Allen, G. Cakins, and J. Helbring. *An ABC Manager's Primer* (New York, NY: IRWIN, 1993).

Tracy, Michael, and Fred Wiersema. *The Discipline of Market Leaders* (Reading, MA: Addison-Wesley, 1995).

Treadway, Morris. *A Primer on Activity-Based Management: ABM in Utilities: A Process for Managing a Market Driver Business* (Coopers & Lybrand 1995).

Turney, Peter B. B. *Common Cents* (Hillsboro, OR: Cost Technology, 1991).

Walther, Thomas, Henry Johansson, John Dunleavy, and Elizabeth Hjelm. *Reinventing the CFO* (New York: McGraw-Hill, 1997).

White, Timothy S. *The 60-Minute ABC Book* (Bedford, TX: CAM-I, 1997).

ARTICLES

Abinanti, Lawson. "Put OLAP to Work in Your Data Warehouse," *Management Accounting* (October 1996), pp. 54–55.

Ackright, Brad. "KCPL Restructures G.L. to Reflect Activities," *As Easy as ABC: ABC Technologies Newsletter* (Fall 1996).

Adams, Robert, and Carter Ray. "United Technologies' Activity-based Accounting Is Catalyst for Success," *As Easy as ABC: ABC Technologies Newsletter* (Fall 1994).

Ainsworth, Penne. "When Activity-based Costing Works," *Practical Accountant* (July 1994), pp. 28–36.

Akright, Brad. "KCPL Gets a Charge Out of ABC/ABM," *As Easy as ABC: ABC Technologies Newsletter* (Winter 1995).

Alkinson, Antony A., John H. Waterhouse, and Robert B. Wells. "A Stakeholder Approach to Strategic Performance Measurement," *Sloan Management Review* (Spring 1997), pp. 25–37.

SUGGESTED READINGS

Anthes, Gary H. "Learning How to Share," *Computerworld,* February 23, 1998, pp. 75–77.

Anthes, Gary H. "The Long Arm of Moore's Law," *Computerworld,* October 5, 1998, p. 69.

Aronhime, Lawrence. "Growing Plants, Cutting Costs: ABC Branches Out into the Nursery Industry," *As Easy as ABC: ABC Technologies Newsletter* (Summer 1994).

Atre, Shaku. "From Build to Buy," *Computerworld,* February 9, 1998, pp. 68–69.

Atre, Shaku. "Learn the Risks of Marts," *Computerworld,* May 19, 1998, pp. 63–64.

Atre, Shaku. "Plan for Data Marts," *Computerworld,* June 16, 1997, pp. 71–72.

Balachandran, Bala. "Cost Management at Saturn: A Case Study," *BusinessWeek Executive Briefing Services* Vol. 5, 1994, pp. 25–28.

Bank, David. "Software Firms Look Outside Windows to Handle Data," *Wall Street Journal,* June 27, 1997, p. B4.

Belli, John. "New Trends in the ERP Selection Process," *Midrange ERP* (September/October 1997), pp. 24–26.

Benjamin, Robert I., and Elliot Levinson. "A Framework for Managing IT-Enabled Change," *Sloan Management Review* (Summer 1993), pp. 23–33.

Bennett, P. "Compumotor's Printed Circuit Assembly," *As Easy as ABC: ABC Technologies Newsletter* (Summer 1996).

Berry, Leonard L., and L. Parasuraman. "Listening to the Customer—The Concept of a Service-Quality Information System," *Sloan Management Review* (Spring 1997), pp. 65–76.

Boltriell, John, and Nabil Aboutanus. "Parks Canada Takes a Rapid Route to Implementing Activity-based Costing," *As Easy as ABC: ABC Technologies Newsletter* (Spring 1997).

Borthick, Fay A., and Harold Roth. "Faster Access to More Information for Better Decisions," *Journal of Cost Management* (Winter 1997), p. 25.

Brausch, John. "Beyond ABC: Target Costing for Profit Enhancement," *Management Accounting* (November 1994), p. 45.

Callahan, V. Charles, and Joseph Nemec. "The CEOs Information Agenda—Seizing the Right Opportunities," *Strategy & Business,* No. 7 (1997), pp. 4–8.

Carlson, David A., and S. Mark Young. "Activity-based Total Quality Management at American Express," *Cost Management* (Spring 1993).

Casper, Carol. "A Value for the Value Chain," *Food Logistics* (June/July 1997), pp. 30–32.

SUGGESTED READINGS

Casper, Carol. "ABC in the Field," *Food Logistics* (June/July 1997), pp. 34–36.

Champy, Jim. "IS and Line Managers Need to Close the Gap," *Computerworld*, January 27, 1977, p. 76.

Chernin, Philip. "ALZA: The First Stages of a Costing Model for Pharmaceutical Manufacturing," *As Easy as ABC: ABC Technologies Newsletter* (Fall 1991).

Cooper, Robin, and Robert S. Kaplan. "Measure Costs Right: Make the Right Decision," *Harvard Business Review* (September-October 1998).

Cooper, Robin, and Robert S. Kaplan. "The Promise and Peril of Integrated Cost Systems," *Harvard Business Review* (July-August 1998), pp. 109–119.

Cooper, Robin. "Squeeze Play," *Journal of Accountancy* (January 1997), pp. 46–48.

Davenport, Thomas D. "Information Behavior: Why We Build Systems that Users Won't Use," *Computerworld*, September 15, 1997.

Davenport, Thomas D. "Putting the Enterprise into the Enterprise System," *Harvard Business Review* (July-August 1998), pp. 121–131.

De Santa, Richard. "Real-life ECR—A Most Measured Approach," *Supermarket Business* (November 1996).

Dedera, Christopher. "Harris Semiconductor ABC: Worldwide Implementation and Total Integration," *Journal of Cost Management* (Spring 1996), pp. 44–58.

Dietze, Clint. "Turning By-products into Dollars," *As Easy as ABC: ABC Technologies Newsletter* (Spring 1995).

Donnelly, J., and Dave Bucanan. "US Airways Takes Off with ABC," *As Easy as ABC: ABC Technologies Newsletter* (Winter 1997).

Downes, Frank A. "Getting the Most Mileage from Your Cost System," *Controller Magazine* (June 1996), pp. 49–53.

Drucker, Peter. "The Information Executives Truly Need," *Harvard Business Review* (January-February 1995), pp. 54–62.

Faulds, John. "Innovative CityMax Turns to ABC to Define Its Future Business," *As Easy as ABC: ABC Technologies Newsletter* (Summer 1997).

Felix, Joanne, and Al Rossman. "Dun & Bradstreet Shared Transaction Services Center: A Case Study in Activity-Based Management," *As Easy as ABC: ABC Technologies Newsletter* (Winter 1995).

Finkelstein, Richard. "MDD: Database Reaches the Next Dimension," *Database Programming & Design*, Vol. 8, No. 4 (April 1995).

Fishman, Charles. "Change," *Fast Company* (April-May 1997), pp. 64–75.

Foley, John. "OLAP Spread," *Industry Week*, October 20, 1997, pp. 20–22.

Frank, Maurice. "A Drill-Down Analysis of Multidimensional Databases," *DBMS* (July 1994), pp. 60–71.

Fraser, F., Dr. Agatha, P.C. IP, and Joseph Young. "Competitive Advantage and Activity-Based Costing: A Hong Kong Case Study," *As Easy as ABC: ABC Technologies Newsletter* (Fall 1996).

Garry, Michael. "ABC in Action," *Progressive Grocer* (February 1996), pp. 71–72.

Geishecker, Mary Lee. "New Technologies Support ABC," *Management Accounting* (March 1996), pp. 42–48.

Gow, Kathleen, and Tom Duffy. "The Support Burden," *Computerworld Global Innovators Series,* June 9, 1997, pp. 9–16.

Harper, Brook, and Ed Trzcienski. "Management Tools Ensuring Quality Service at Knight-Ridder's Shared Services Center," *As Easy as ABC: ABC Technologies Newsletter* (Fall 1996).

Haynes, Peter. "The Computer Industry," *The Economist,* September 17, 1994.

Hermance, Dave. "A Shift in Business Philosophy: AMETEK Moves to ABM," *As Easy as ABC: ABC Technologies Newsletter* (Fall 1994).

Hibbard, Justin. "Knowledge Management: Knowing What We Know," *Industry Week,* October 20, 1997, pp. 46–178.

Hildedrand, Carol. "Call Weighting," *CIO,* February 15, 1998, pp. 54–59.

Hoffman, Thomas. "Datawarehouse, the Sequel," *Computerworld,* June 2, 1997, pp. 69–72.

Hoffman, Thomas. "Poor Profit Data Weakens Business," *Computerworld,* April 6, 1998, p. 39.

Howard, Patrick. "Architecture for an Activity-Based Costing System," *Journal of Cost Management* (Winter 1995).

Hubbell, William W. "Combining Economic-Value-Added and Activity-based Management," *Journal of Cost Management* (Spring 1996), pp. 18–30.

Inmon, W. H. "The Data Warehouse: All Your Data at Your Fingertips," *Communications Week,* August 29, 1994.

Jessup, S., and D. Stinchfield. "Homecare Agency Counts ABC as a Key Strategic Tool," *As Easy as ABC: ABC Technologies Newsletter* (Spring 1996).

Lazare, Cathy. "Giant Steps in ABM," *Controller Magazine* (September 1997), pp. 39–44.

Leahy, Tad. "ABM—When to Hold 'em, When to Fold 'em," *Business Finance* (November 1998), pp. 53–54.

Leahy, Tad. "Beyond Traditional Product Costing," *Controller Magazine* (August 1998), pp. 65–68.

Leahy, Tad. "Making Sure the Customer Is Always Bright," *Business Finance* (October 1998), pp. 61–63.

LeBlanc, J., and Tom Roberts. "ABC Assists in Building a More Profitable CASE," *As Easy as ABC: ABC Technologies Newsletter* (Fall 1997).

SUGGESTED READINGS

Lev, Baruch. "The Old Rules No Longer Apply," *Forbes ASAP,* April 7, 1997, pp. 35–36.

Lyons, Lawrence S. "Creating Tomorrow's Organization: Unlocking the Benefits of Future Work," *Leader to Leader* (Summer 1997), p. 9.

Mangan, Thomas N. "Integrating and Activity-based Cost System," *Journal of Cost Management* (Winter 1995).

Maxim, Jim. "No More Silver Bullets," *Industry Week,* September 1, 1997, pp. 42–43.

Mckie, Stewart. "The Power of OLAP," *Controller Magazine* (January 1997), pp. 34–39.

Mckie, Stewart. "Mining Your Accounting Data," *Controller Magazine* (November 1996), pp. 43–46.

McLemore, Ivy. "Overhauling Cost Accounting Systems," *Controller Magazine* (November 1996), pp. 43–46.

McNair, C. J. "To Serve the Customer Within," *Journal of Cost Management* (Winter 1996), pp. 40–43.

Mcvitty, Amanda. "Decision-making Made Easy," *Management Technical Briefing* (June 1998), pp. 27–32.

Merryman, Jim. "Food Industry Leader Uses ABC Recipe for Success," *As Easy as ABC: ABC Technologies Newsletter* (Summer 1991).

Meyer, Christopher. "How the Right Measures Help Teams Excel," *Harvard Business Review* (May-June 1994), pp. 95–103.

Miller, Don, Randall Benson, and Rick Sahli. "Providence Portland Medical Center Gets a $1.6 M Shot in the Arm," *As Easy as ABC: ABC Technologies Newsletter* (Fall 1997).

More E., Jorge Medina. "Grupo Casa Autrey's CFO Drives Profitability Using ABC," *As Easy as ABC: ABC Technologies Newsletter* (Summer 1998).

Morris, Henry. "ABC Technologies: A Business Methodology Foundation for Analytic Applications," *IDC Bulletin,* No. 16838 (August 1998).

Morris, Henry. "Applications and Information Access: Information Access Tools," *IDC Bulletin,* No. 14064 (August 1997).

Morris, Henry. "Packaging the Vertical Warehouse: Cognos' Application Partnerships," *IDC Bulletin,* No. 18340, An International Data Corp. White Paper (November 1996).

Morrissey, Eileen, and Gary Hodson. "A Smarter Way to Run a Business," *Journal of Accountancy* (January 1997), pp. 48–50.

Nair, Mohan. "Learning Your ABC," *The Hong Kong Accountant Journal,* FT Law & Tax Hong Kong—Pierson Publishing.

Norkiewicz, Angela. "Nine Steps to Implementing ABC," *Management Accounting* (April 1994).

Panchak, Patricia. "Manufacturing—the Future," *Industry Week,* September 21, 1998, pp. 97–105.

Parker, Robert. "If It's Not Broke, Break It," *Controller Magazine* (June 1997), pp. 67–68.

Patras, Dan, and Kevin Clancy. "ABC in the Service Industry: Product Line Profitability at Acordia, Inc." *As Easy as ABC: ABC Technologies Newsletter* (Spring 1993).

Payne, Fred. "Lloyds Bank Uses ABM to Expand Service Volume and Reduce Costs," *As Easy as ABC: ABC Technologies Newsletter* (Spring 1997).

Perry, R. A., and J. R. Spaulding. "The Future Is Now: Rolling Out ABC at the Valvoline Company," *As Easy as ABC: ABC Technologies Newsletter* (Spring 1994).

Perry, Dick, and John Spaulding. "An ABC-case Study—Valvoline," *ABC Technologies* 1994.

Phillips, K., and Kevin Dilton-Hill. "Willards Foods: Managing Customer Profitability with ABC Information," *As Easy as ABC: ABC Technologies Newsletter* (Winter 1996).

Player, Steve. "The ABM Tidal Wave," *Controller Magazine* (December 1997), pp. 71–72.

Player, Steve. "Seven Key Project Aids for Global Services," *Cost Management Trends* (October 1997).

Player, Steve. "Solving the Customer Profitability Puzzle," *Controller Magazine* (June 1992), pp. 65–66.

Porter, Michael. "What Is Strategy," *Harvard Business Review* (November-December 1996), pp. 61–78.

Rautio, Jan. "Activity Analysis: Key to Service Industry Success," *As Easy as ABC: ABC Technologies Newsletter* (Winter 1991).

Ruber, Peter. "EIS: Selling to the Top," *VarBusiness* (October 1994), p. 91.

SAM Project Team. "Hewlett Packard Knows What It Takes and What It Costs," *As Easy as ABC: ABC Technologies Newsletter* (Summer 1995).

Sammer, Joanne. "Customer-driven Activity-based Costing Software Solutions," *Controller Magazine* (November 1997), pp. 69–76.

Schaefer, S. "ABC Implementation: Planning Is Everything," *As Easy as ABC: ABC Technologies Newsletter* (Spring 1991).

Schiemann, William, and John Lingle. "Seven Greatest Myths of Measurement," *Management Review* (May 1997), p. 29.

Schmitt, Thomas. "ABC and ECR: Not Just Alphabet Soup," *As Easy as ABC: ABC Technologies Newsletter* (Winter 1995).

Shaw, Russell. "ABC and ERP: Partners at Last?" *Management Accounting* (November 1998), pp. 56–58.

SUGGESTED READINGS

Spensky, Stuart. "Coats Viyella Weaves a New Business Model," *As Easy as ABC: ABC Technologies Newsletter* (Summer 1996).

Stedmand, Craig. "Augmenting Data Depots," *Computerworld,* June 16, 1997, pp. 57–60.

Stedmand, Craig. "Do Users Know Your Data?" *Computerworld,* August 11, 1997, pp. 37–40.

Taninecz, George. "All for One Enterprise," *Industry Week,* September 1, 1997, pp. 30–36.

Teach, Edward. "View Masters," *CFO* (January 1997), pp. 47–52.

Thebes, Thomas H., Jr., and Steve Millward. "Exide Electronics' Large Systems Group Tackles Multi-dimensional Manufacturing," *As Easy as ABC: ABC Technologies Newsletter* (Summer 1995).

Tyre, Marcie J., and Wanda J. Orlikowski. "Exploiting Opportunities for Technological Improvement in Organizations," *Sloan Management Review* (Fall 1993), p. 13.

Vadgama, Ashok. "Using Activity-Based Management for Project Management," *As Easy as ABC: ABC Technologies Newsletter* (Summer 1996).

Kim, W. Chan and Renee Mauborgne. "Value Innovation: The Strategic Logic of High Growth," *Harvard Business Review* (January-February 1997), pp. 103–112.

Walkington, Dagmar. "Understanding the Cost of Post Processes and Products—The New Zealand Post Experience," *As Easy as ABC: ABC Technologies Newsletter* (Summer 1997).

Walter, Jeff, Pat Raltan, Steve Smith, and Doug Webster. "ABC at NASA's Lewis Research Center," *As Easy as ABC: ABC Technologies Newsletter* (Winter 1995).

Whitney, John. "Strategic Renewal for Business Units," *Harvard Business Review* (July-August 1996), pp. 84–98.

Williamson, Miryam. "From SAP to NUTS," *Computerworld,* November 10, 1997, pp. 68–69.

Willis, Rich. "Major Boo-Boo," *Forbes ASAP,* April 7, 1997, p. 36.

Xenakis, John. "Software for All Reasons," *CFO* (February 1997), pp. 53–68.

Young, Debby. "Score It a Hit," *CIO Enterprise,* November 15, 1998, p. 27.

Zones, Aaron. "Building a Business Case," *Communications Week,* August 29, 1994.

Zuberi, Faheem. "Sallie Mae Puts ABC to the Test," *As Easy as ABC: ABC Technologies Newsletter* (Spring 1998).

Zweben, Monte. "Beyond ERP: Enterprise Resource Optimization," *Midrange ERP* (September/October 1997), pp. 50–54.

CAM-I GLOSSARY OF TERMS*

COMMON TERMINOLOGY

By the time this glossary was assembled, many competing terms were already in use. Given that several individuals and organizations were doing work in the area of Activity-Based Costing, it is not surprising to find identical terms with different meanings. It was one of the objectives of this project to compile a set of terms and definitions that would provide a single data base for the CAM-I CMS Program and others working in this area. We realize that this project in itself may add to the diversity of terms. However, it is essential that if the CAM-I CMS Program is to continue to develop and publish work on the theory and practice of Activity-Based Management, we need a single and accepted set of terms and definitions.

This issue became very clear when we looked at the variations of the term driver in use, as an example. It is the intent of this glossary to foster an industry standard for the terminology of this discipline. The work by others in this field should be fully recognized, acknowledged, and appreciated. While it is the intent of this glossary to fulfill the need of CAM-I sponsor members who require a common set of definitions that will permit them to communicate effectively with each other, in the longer view, an industry standard is far more preferable. If interested parties outside of CAM-I wish to adopt the terms and definitions contained in this work, we would consider that to be a positive comment on our work. We would also be glad to accept comment from outside of CAM-I on this work so that it can be continually improved as new editions are published.

APPENDICES

Contained in this glossary are several appendices. Due to the early stage of maturity and evolution of the Activity-Based Management, it was decided that several topics should be expanded in order to share our views on key issues. We hope that these additional sections assist in explaining why certain terms were chosen and the context in which we perceive their use.

---

* Used with permission. Norm Raffish and Peter B. B. Turney, (eds.), *The CAM-I Glossary of Activity-Based Management, Version 1.2* (Arlington, TX: The Consortium for Advanced Manufacturing-International, 1992).

GLOSSARY OF TERMS

ABC See *activity-based costing.*

Absorption costing A method of costing that assigns all or a portion of the manufacturing costs to products or other cost objects. The costs assigned include those that vary with the level of activity performed and also those that do not vary with the level of activity performed.

Activity 1. Work performed within an organization. 2. The aggregations of actions performed within an organization that are useful for purposes of activity-based costing.

Activity analysis The identification and description of activities in an organization. Activity analysis involves determining what activities are done within a department, how many people perform the activities, how much time they spend performing the activities, what resources are required to perform the activities, what operational data best reflect the performance of the activities, and what value the activity has for the organization. Activity analysis is accomplished by means of interviews, questionnaires, observations, and reviews of physical records of work.

Activity attributes Characteristics of individual activities. Attributes include cost drivers, cycle time, capacity, and performance measures. For example, a measure of the elapsed time required to complete an activity is an attribute (See *cost driver* and *performance measures.*)

Activity capacity The demonstrated or expected capacity of an activity under normal operating conditions, assuming a specified set of resources and over a long period of time. An example of this would be a rate of output for an activity expressed as 500 cycles per hour.

Activity cost assignment The process in which the cost of activities are attached to cost objects using activity drivers. (See *cost object* and *activity driver.*)

Activity cost pool A grouping of all cost elements associated with an activity. (See *cost element.*)

Activity driver A measure of the frequency and intensity of the demands placed on activities by cost objects. An activity driver is used to assign costs to cost objects. It represents a line-item on the bill of activities for a product or customer. An example is the number of part numbers, which is used to measure the consumption of material-related activities by each product, material type, or component. The number of customer orders measures the consumption of order-entry activities by each customer. Sometimes an activity driver is used as an indicator of the output of an activity, such as the number of purchase orders prepared by the purchasing activity. (See *intensity, cost object,* and *bill of activities.*)

CAM-I GLOSSARY OF TERMS

Activity driver analysis The identification and evaluation of the activity drivers used to trace the cost of activities to cost objects. Activity driver analysis may also involve selecting activity drivers with a potential for cost reduction. (See *Pareto analysis*.)

Activity level A description of how an activity is used by a cost object or other activity. Some activity levels describe the cost object that uses the activity and the nature of this use. These levels include activities that are traceable to the product (i.e., unit-level, batch-level, and product-level costs), to the customer (customer-level costs), to a market (market-level costs), to a distribution channel (channel-level costs), and to a project, such as a research and development project (project-level costs).

Activity-based costing A methodology that measures the cost and performance of activities, resources, and cost objects. Resources are assigned to activities, then activities are assigned to cost objects based on their use. Activity-based costing recognizes the causal relationships of cost drivers to activities.

Activity-based cost system A system that maintains and processes financial and operating data on a firm's resources, activities, cost objects, cost drivers, and activity performance measures. It also assigns cost to activities and cost objects.

Activity-based management A discipline that focuses on the management of activities as the route to improving the value received by the customer and the profit achieved by providing this value. This discipline includes cost driver analysis, activity analysis, and performance measurement. Activity-based management draws on Activity-based Costing as its major source of information. (See *customer value*.)

Allocation 1. An apportionment or distribution. 2. A process of assigning cost to an activity or cost object when a direct measure does not exist. For example, assigning the cost of power to a machine activity by means of machine hours is an allocation, because machine hours is an indirect measure of power consumption. In some cases, allocations can be converted to tracing by incurring additional measurement costs. Instead of using machine hours to allocate power consumption, for example, a company can place a power meter on machines to measure actual power consumption. (See *tracing*.)

Assignment See *cost assignment*.

Attributes Characteristics of activities, such as cost drivers and performance measures. (See *cost driver* and *performance measure*.)

Attribution See *tracing*.

Avoidable cost A cost associated with an activity that would not be incurred if the activity was not required. The telephone cost associated with vendor support, for example, could be avoided if the activity were not performed.

Backflush costing   1. A costing method that applies costs based on the output of a process. The process uses a bill of material or a bill of activities explosion to draw quantities from inventory, through work-in-process, to finished goods; at any intermediate stage, using the output quantity as the basis. These quantities are generally costed using standard costs. The process assumes that the bill of material (or bill of activities) and the standard costs at the time of backflushing represent the actual quantities and resources used in the manufacture of the product. This is important, since no shop orders are usually maintained to collect costs. 2. A costing method generally associated with repetitive manufacturing. (See *repetitive manufacturing* and *standard costing*.)

Benchmarking   See *best practices.*

Best Practices   A methodology that identifies an activity as the benchmark by which a similar activity will be judged. This methodology is used to assist in identifying a process or technique that can increase the effectiveness or efficiency of an activity. The source may be internal (e.g., taken from another part of the company) or external (e.g., taken from a competitor.) Another term used is *competitive benchmarking.*

Bill of activities   A listing of the activities required (and, optionally, the associated costs of the resources consumed) by a product or other cost object.

Budget   1. A projected amount of cost or revenue for an activity or organizational unit covering a specific period of time. 2. Any plan for the coordination and control of resources and expenditures.

Capital decay   1. A quantification of the lost revenues or reduction in net cash flows sustained by an entity due to obsolete technology. 2. A measure of uncompetitiveness.

Carrying cost   See *holding cost.*

Competitive benchmarking   See *best practices.*

Continuous improvement program   A program to eliminate waste, reduce response time, simplify the design of both products and processes, and improve quality.

Cost Accounting Standards   1. Rules promulgated by the Cost Accounting Standards Board of the United States Government to ensure contractor compliance in the accounting of government contracts. 2. A set of rules issued by any of several authorized organizations or agencies, such as the American Institute of Certified Public Accountants (AICPA) or the Association of Chartered Accountants (ACA), dealing with the determination of costs to be allocated, inventoried, or expensed.

Cost assignment   The tracing or allocation of resources to activities or cost objects. (See *allocation* and *tracing*.)

Cost center   The basic unit of responsibility in an organization for which costs are accumulated.

CAM-I GLOSSARY OF TERMS

Cost driver Any factor that causes a change in the cost of an activity. For example, the quality of parts received by an activity (e.g., the percent that are defective) is a determining factor in the work required by that activity, because the quality of parts received affects the resources required to perform the activity. An activity may have multiple cost drivers associated with it.

Cost driver analysis The examination, quantification, and explanation of the effects of cost drivers. Management often uses the results of cost driver analyses in continuous improvement programs to help reduce throughput time, improve quality, and reduce costs. (See *cost driver* and *continuous improvement program.*)

Cost element An amount paid for a resource consumed by an activity and included in an activity cost pool. For example, power cost, engineering cost, and depreciation may be cost elements in the activity cost pool for a machine activity. (See *activity cost pool, bill of activities,* and *resource.*)

Cost object Any customer, product, service, contract, project, or other work unit for which a separate cost measurement is desired.

Cost of quality All the resources expended for appraisal costs, prevention costs, and both internal and external failure costs of activities and cost objects.

Cost pool See *activity cost pool.*

Cross-subsidy The improper assignment of costs among cost objects such that certain cost objects are overcosted while other cost objects are under-costed relative to the activity costs assigned. For example, traditional cost accounting systems tend to overcost high-volume products and undercost low-volume products.

Customer value The difference between customer realization and sacrifice. *Realization* is what the customer receives, which includes product features, quality, and service. This takes into account the customer's cost to use, maintain, and dispose of the product or service. *Sacrifice* is what the customer gives up, which includes the amount the customer pays for the product plus time and effort spent acquiring the product and learning how to use it. Maximizing customer value means maximizing the difference between realization and sacrifice.

Differential cost See *incremental cost.*

Direct cost A cost that is traced directly to an activity or a cost object. For example, the material issued to a particular work order or the engineering time devoted to a specific product are direct costs to the work orders or products. (See *tracing.*)

Direct tracing See *tracing.*

Discounted cash flow A technique used to evaluate the future cash flows generated by a capital investment. Discounted cash flow is computed by discounting cash flows to determine their present value.

Diversity   Conditions in which cost objects place different demands on activities or activities place different demands on resources. This situation arises, for example, when there is a difference in mix or volume of products that causes an uneven assignment of costs. Different types of diversity include: batch size, customer, market, product mix, distribution channel, and volume.

Financial accounting   1. The accounting for assets, liabilities, equities, revenues, and expenses as a basis for reports to external parties. 2. A methodology that focuses on reporting financial information primarily for use by owners, external organizations, financial institutions. This methodology is constrained by rule-making bodies such as the Financial Accounting Standards Board (FASB), the Securities Exchange Commission (SEC), and the American Institute of Certified Public Accountants (AICPA).

First-stage allocation   See *resource cost assignment.*

Fixed cost   A cost element of an activity that does not vary with changes in the volume of cost drivers or activity drivers. The depreciation of a machine, for example, may be direct to a particular activity, but it is fixed with respect to changes in the number of units of the activity driver. The designation of a cost element as fixed or variable may vary depending on the time frame of the decision in question and the extent to which the volume of production, activity drivers, or cost drivers changes.

Flexible factory   The objective of a flexible factory is to provide a wide range of services across many product lines in a timely manner. An example is a fabrication plant with several integrated manufacturing cells that can perform many functions for unrelated product lines with relatively short lead times.

Focused factory   The objective of a focused factory is to organize around a specific set of resources to provide low cost and high throughput over a narrow range of products.

Forcing   Allocating the costs of a sustaining activity to a cost object even though that cost object may not clearly consume or causally relate to that activity. Allocating a plant-level activity (such as heating) to product units using an activity driver such as direct labor hours, for example, forces the cost of this activity to the product. (See *sustaining activity.*)

Full absorption costing   See *absorption costing.*

Functional decomposition   Identifies the activities performed in the organization. It yields a hierarchical representation of the organization and shows the relationship between the different levels of the organization and its activities. For example, a hierarchy may start with the division and move down through the plant, function, process, activity, and task levels.

Holding cost   A financial technique that calculates the cost of retaining an asset (e.g., finished goods inventory or a building). Generally, the calculation includes a cost of capital in addition to other costs such as insurance, taxes, and space.

Homogeneity A situation in which all the cost elements in an activity's cost pool are consumed in proportion to an activity driver by all cost objects. (See *cost element, activity cost pool,* and *activity driver.*)

Incremental cost 1. The cost associated with increasing the output of an activity or project above some base level. 2. The additional cost associated with selecting one economic or business alternative over another, such as the difference between working overtime or subcontracting the work. 3. The cost associated with increasing the quantity of a cost driver. (Also known as *differential cost.*)

Indirect cost The cost that is allocated—as opposed to being traced—to an activity or a cost object. For example, the costs of supervision or heat may be allocated to an activity on the basis of direct labor hours. (See *allocation.*)

Intensity The cost consumed by each unit of the activity driver. It is assumed that the intensity of each unit of the activity driver for a single activity is equal. Unequal intensity means that the activity should be broken into smaller activities or that a different activity driver should be chosen. (See *diversity.*)

Life Cycle See *product life cycle.*

Net present value A method that evaluates the difference between the present value of all cash inflows and outflows of an investment using a given rate of discount. If the discounted cash inflow exceeds the discounted outflow, the investment is considered economically feasible.

Non–value-added activity An activity that is considered not to contribute to customer value or to the organization's needs. The designation non–value-added reflects a belief that the activity can be redesigned, reduced, or eliminated without reducing the quantity, responsiveness, or quality of the output required by the customer or the organization. (See *customer value* and *value analysis.*)

Obsolescence A product or service that has lost its value to the customer due to changes in need or technology.

Opportunity cost The economic value of a benefit that is sacrificed when an alternative course of action is selected.

Pareto analysis The identification and interpretation of significant factors using Pareto's rule that 20 percent of a set of independent variables is responsible for 80 percent of the result. Pareto analysis can be used to identify cost drivers or activity drivers that are responsible for the majority of cost incurred by ranking the cost drivers in order of value. (See *cost driver analysis* and *activity driver analysis.*)

Performance measures Indicators of the work performed and the results achieved in an activity, process, or organizational unit. Performance measures may be financial or nonfinancial. An example of a performance measure of an activity is the number of defective parts per million. An example of a performance measure of an organizational unit is return on sales.

Present value The discounted value of a future sum or stream of cash flows.

Process A series of activities that are linked to perform a specific objective. For example, the assembly of a television set or the paying of a bill or claim entails several linked activities.

Product family A group of products or services that have a defined relationship because of physical and production similarities. (The term *product line* is used interchangeably.)

Product life cycle The period that starts with the initial product specification and ends with the withdrawal of the product from the marketplace. A product life cycle is characterized by certain defined stages, including research, development, introduction, maturity, decline, and abandonment.

Product line See *product family*.

Profit center A segment of the business (e.g., a project, program, or business unit) that is accountable for both revenues and expenses.

Project A planned undertaking, usually related to a specific activity, such as the research and development of a new product or the redesign of the layout of a plant.

Project costing A cost system that collects information on activities and costs associated with a specific activity, project, or program.

Repetitive manufacturing The manufacture of identical products (or a family of products) in a continuous flow.

Resource An economic element that is applied or used in the performance of activities. Salaries and materials, for example, are resources used in the performance of activities. (See *cost element*.)

Resource cost assignment The process by which cost is attached to activities. This process requires the assignment of cost from general ledger accounts to activities using resource drivers. For example, the chart of accounts may list information services at a plant level. It then becomes necessary to trace (assuming that tracing is practical) or to allocate (when tracing is not practical) the cost of information services to the activities that benefit from the information services by means of appropriate resource drivers. It may be necessary to set up intermediate activity cost pools to accumulate related costs from various resources before the assignment can be made. (See *activity cost pool* and *resource driver*.)

Resource driver A measure of the quantity of resources consumed by an activity. An example of a resource driver is the percentage of total square feet of space occupied by an activity. This factor is used to allocate a portion of the cost of operating the facilities to the activity.

Responsibility accounting   An accounting method that focuses on identifying persons or organizational units that are accountable for the performance of revenue or expense plans.

Risk   The subjective assessment of the possible positive or negative consequences of a current or future action. In a business sense, risk is the premium asked or paid for engaging in an investment or venture. Often risk is incorporated into business decisions through such factors as hurdle rates or the interest premium paid over a prevailing base interest rate.

Second-stage allocation   See *activity cost assignment.*

Standard costing   A costing method that attaches costs to cost objects based on reasonable estimates or cost studies and by means of budgeted rates rather according to actual costs incurred.

Sunk costs   Costs that have been invested in assets for which there is little (if any) alternative or continued value except salvage. Using sunk costs as a basis for evaluating alternatives may lead to incorrect decisions. Examples are the invested cost in a scrapped part or the cost of an obsolete machine.

Support costs   Costs of activities not directly associated with production. Examples are the costs of process engineering and purchasing.

Surrogate activity driver   An activity driver that is not descriptive of an activity, but that is closely correlated to the performance of the activity. The use of a surrogate activity driver should reduce measurement costs without significantly increasing the costing bias. The number of production runs, for example, is not descriptive of the material disbursing activity, but the number of production runs may be used as an activity driver if material disbursements coincide with production runs.

Sustaining activity   An activity that benefits an organization at some level (e.g., the company as a whole or a division, plant, or department), but not any specific cost object. Examples of such activities are preparation of financial statements, plant management, and the support of community programs.

Target cost   A cost calculated by subtracting a desired profit margin from an estimated (or a market-based) price to arrive at a desired production, engineering, or marketing cost. The target cost may not be the initial production cost, but instead the cost that is expected to be achieved during the mature production stage. (See *target costing.*)

Target costing   A method used in the analysis of product and process design that involves estimating a target cost and designing the product to meet that cost. (See *target cost.*)

Technology costs   A category of cost associated with the development, acquisition, implementation, and maintenance of technology assets. It can include costs such as the depreciation of research equipment, tooling amortization, maintenance, and software development.

Technology valuation A nontraditional approach to valuing technology acquisitions that may incorporate such elements as purchase price, start-up costs, current market value adjustments, and the risk premium of an acquisition.

Throughput The rate of production of a defined process over a stated period of time. Rates may be expressed in terms of units of products, batches produced, dollar turnover, or other meaningful measurements.

Traceability The ability to assign a cost by means of a causal relationship directly to an activity or a cost object in an economically feasible way. (See *tracing*.)

Tracing The assignment of cost to an activity or a cost object using an observable measure of the consumption of resources by the activity or cost object. Tracing is generally preferred to allocation if the data exist or can be obtained at a reasonable cost. For example, if a company's cost accounting system captures the cost of supplies according to which activity uses the supplies, the costs may be traced—as opposed to allocated—to the appropriate activities. Tracing is also called *direct tracing*.

Unit cost The cost associated with a single unit of the product, including direct costs, indirect costs, traced costs, and allocated costs.

Value-added activity An activity that is judged to contribute to customer value or satisfy an organizational need. The attribute "value-added" reflects a belief that the activity cannot be eliminated without reducing the quantity, responsiveness, or quality of output required by a customer or organization. (See *customer value*.)

Value analysis A cost reduction and process improvement tool that utilizes information collected about business processes and examines various attributes of the processes (e.g., diversity, capacity, and complexity) to identify candidates for improvement efforts. (See *activity attribute* and *cost driver*.)

Value chain The set of activities required to design, procure, produce, market, distribute, and service a product or service.

Value-chain costing An activity-based cost model that contains all activities in the value chain.

Variance The difference between an expected and actual result.

Variable cost A cost element of an activity that varies with changes in volume of cost drivers and activity drivers. The cost of material handling to an activity, for example, varies according to the number of material deliveries and pickups to and from that activity. (See *cost element, fixed cost,* and *activity driver*.)

Waste Resources consumed by unessential or inefficient activities.

Willie Sutton rule Focus on the high-cost activities. The rule is named after bank robber Willie Sutton, who—when asked "why do you rob banks?"—is reputed to have replied "because that's where the money is."

Work cell A physical or logical grouping of resources that performs a defined job or task. The work cell may contain more than one activity. For example, all the tasks associated with the final assembly of a product may be grouped in a work cell.

Work center A physical area of the plant or factory. It consists of one or more resources where a particular product or process is accomplished.

APPENDIX A

Choice of Terms

Driver
There is probably no term, other than activity, that has become more identified with Activity-Based Costing as the term driver and its several variations. The problem is that it has been applied to several entities with varying meanings. It is often difficult to understand whether the use of the term driver is related to a causal effect (cost or input driver) or to the output of an activity (cost or output driver). In addition, terms such as first and second stage driver have come into use which also describe entities similar to resource and activity driver.

In this glossary we have chosen to use the term cost driver as the causal event that influences the quantity of work, and therefore costs, in an activity. We believe that by restricting the definition of cost driver to one meaning, it will facilitate its understanding.

We also appended the term driver to two other entities. The first deals with the mechanism of assigning resources to activities. That we call an resource driver. The second deals with the mechanism of assigning activity costs to cost objects. That we call an activity driver.

We hope that by limiting the use of the word driver to three clearly defined entities, we can prevent misinterpretation or misuse of the term.

Non–Value-Added and Sustaining Activities
There are many activities in an organization that do not contribute to customer value, responsiveness, and quality.

APPENDIX B

Illustrations

The CAM-I ABC Basic Model
The first illustration of the basic model is an attempt to establish a generic illustration that can be used to assist in explaining the concepts of Activity-Based Costing. The model should be thought of as a template that can be

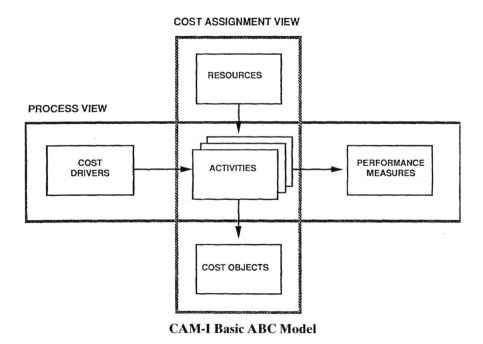

CAM-I Basic ABC Model

adapted for various purposes. The model should not be thought of as a flow chart of activity-based costing. It is meant to be a conceptual diagram that allows the reader to gain a high level understanding of the ABC processes.

There are two axes to the model. The vertical one deals with the classic two-stage cost assignment view. In the expanded model, their are three entities and two processes. The resource entity contains all available means upon which the selected activity can draw. The resource cost assignment process contains the structure and tools to trace and allocate costs to the activity. It is during this process that the applicable resource drivers are developed as the mechanism to convey resource costs to the activity. The activity entity is where work is performed. In this view, the activity is part of the cost structure. It is where resources are converted to some type of output. The activity cost assignment process contains the structure and tools to assign costs to cost objects, utilizing activity drivers as the mechanism to accomplish this assignment.

This cost assignment view is basically a "snap-shot" view in the sense that the Balance Sheet on a financial statement is only a view of the business at the moment the accounts were tallied. In this sense, the cost assignment view can be seen as the structure and rules by which cost assignment takes place at some specific time. This time period may be at the end of a month, quarter, or any other time period which may or may not coincide with an accounting reporting period.

The horizontal axis contains the process view. This is a dynamic view, similar to the Income and Expense statement that reports on what has/is happening. This part of the process is initiated by a causal occurrence we call a cost driver. The cost driver is the agent that causes the activity to utilize resources to accomplish some designated work. In this view the activity is some type of active work center. During and after the activity work effort, performance data are collected. The performance measure of activities entity houses the evaluative criteria by which the organization can determine the efficiency and effectiveness of the activities work effort. It should be noted that there are many other performance measures, such as market share and return on equity, that are not included in the performance measures included in the ABC model.

The process view will constantly be changing. Each time a cost driver initiates work in an activity, new results will be obtained. It is therefore critical that applicable and realistic performance measures be established so that tracking of activity results can be monitored and improved on a continuing basis. ABC, through its reporting and analysis, can become an enabler of other process changes such as synchronous manufacturing, Design for Manufacturing, and Design for Assembly.

Expanded Process View
The second illustration is a more realistic view of what really takes place in an organization. There are many processes in progress, and each is usually made up of several linked activities. The illustration points out that the output (cost object) of any activity may be the input (cost driver) of the next activity. This relationship, of several activities forming a process or sub-process, offers the opportunity to link congruent performance measurements which would offer a more appropriate view of the effectiveness and efficiency of that process.

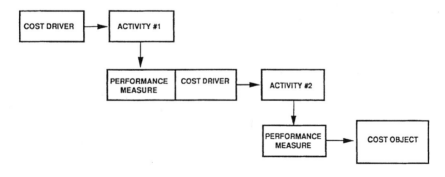

Expanded Process View

CAM-I Expanded ABC Model
The third illustration displays an expanded view of the ABC model. Depicted in this illustration are the resource cost assignment and activity cost assignment processes and their respective data bases of drivers. Another addition is an entity called the activity trigger. This term is not defined in the glossary as it is pertains to an activity-based costing system rather than to the ABC methodology. The activity trigger is often, but not always, the link between the occurrence of a cost driver and the initiation of action in an activity. As an example, the mere occurrence of scrap does not in itself initiate an activity. There will need to be some management authorization to proceed before a replacement part is produced. In an information system about ABC, the activity trigger will often be the collection point for the information about the cost driver. The other entities depicted are cost drivers, activities, cost objects, and activity related performance measures.

ABC Model Example
Included as well in this section is the fourth illustration that displays how the model might be applied to a functional activity. The activity illustrated here is the purchasing activity at a department level. The particular task involved is generating purchase orders. One can see by the metrics selected for perfor-

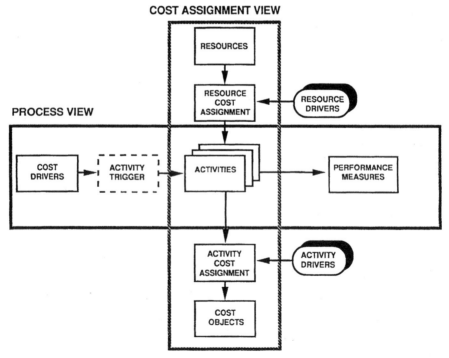

CAM-I Expanded ABC Model

CAM-I GLOSSARY OF TERMS

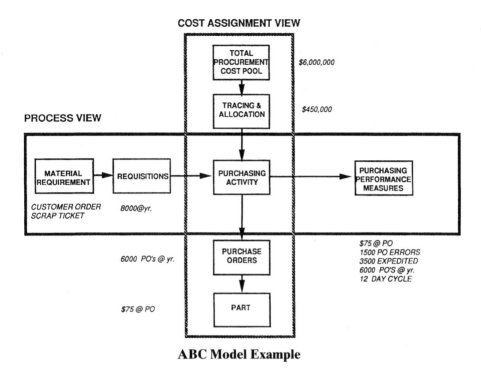

ABC Model Example

mance measures that a trend analysis could certainly identify candidate tasks for a continuous improvement program.

Activity-Based Management Model
The last illustration is a view of Activity-Based Management. It depicts the key relationship between ABC and the management analysis tools that are needed

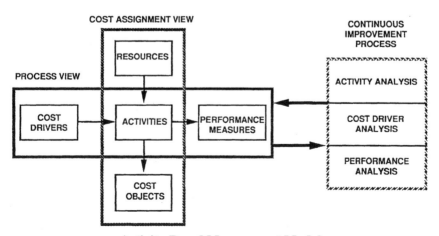

Activity-Based Management Model

to bring full realization of the benefits of ABC to the organization. ABC is a methodology that can yield significant information about cost drivers, activities, resources, and performance measures. However, ABM is a discipline that offers the organization the opportunity to improve the value of its products and services.

APPENDIX C

References

In attempting to use as much of what was available in terms of definitions, there were several reference documents that provided excellent source material. They are:

1. *A Dictionary for Accountants* by Eric L. Kohler, Fifth Edition (Englewood Cliffs: Prentice-Hall, 1975)
2. "Management Accounting Terminology, Statement on Management Accounting Number 2A" (National Association of Accountants, May 1990)
3. *Webster's Ninth New Collegiate Dictionary* (Merriam-Webster, Inc., 1984)
4. *Cost Management for Today's Advanced Manufacturing,* Edited by C. Berliner and J. A. Brimson (Harvard Business School Press 1988)

Activity-Based Costing has gained a significant measure of development and publicity in the past few years. Several individuals have made significant contributions to this growing body of knowledge. Through their articles, books and lectures, they have influenced almost all of us who have worked in the field of Activity-Based Management. It is also fair to say that their efforts have influenced the content of this glossary.

1. H. Thomas Johnson and Robert S. Kaplan, *Relevance Lost: The Rise and Fall of Management Accounting,* (Harvard Business Review, 1987)
2. Thomas Johnson, a series of articles:

    Activity Management: Reviewing the Past and Future of Cost Management (*Journal of Cost Management,* Winter 1990)

    Pitfalls in Using ABC Cost-Driver Information to Manage Operating Cost (*Corporate Controller,* January–February 1991, coauthors T. P. Vance and R. S. Player)

    Activity Management: Past, Present, and Future (*The Engineering Economist,* Spring 1991)

3. Robin Cooper, a series of articles:

   Schrader Bellows (*Harvard Business School* Case, 1986)

   The Rise of Activity Based Costing, in four parts (*Journal of Cost Management,* Summer and Fall 1988, Winter 1989 and Spring 1990)

   The Two Stage Procedure in Cost Accounting, in two parts (*Journal of Cost Management,* Spring and Summer 1987)

   Cost Classifications in Unit-Based and Activity-Based Management Cost Systems (*Journal of Cost Management,* Fall 1990)

4. Robert S. Kaplan, Union Pacific, (*Harvard Business School* Case, 1987)
5. Robert S. Kaplan, One Cost System Isn't Enough (*Harvard Business Review,* January-February 1988)
6. Robin Cooper and Robert S. Kaplan, Measure Costs Right: Make the Right Decisions. (*Harvard Business Review,* September-October 1988)
7. Peter B. B. Turney, a series of articles:

   What Is the Scope of Activity-Based Costing (*Journal of Cost Management,* Fall 1988)

   Ten Myths About Implementing Activity-Based Cost Systems (*Journal of Cost Management,* Spring 1990)

   The Impact of Continuous Improvement on the Design of Activity-Based Cost Systems, with James Reeve (*Journal of Cost Management,* Summer 1990)

   How Activity-Based Costing Helps Reduce Cost (*Journal of Cost Management,* Winter 1991)

8. James Brimson, *Activity Accounting: An Activity-Based Cost Approach* (Coopers and Lybrand, John Wiley & Sons, 1991)
9. M. Stahl and G. Bound, editors, *Competing Globally Through Customer Value: The Management of Suprasystems* (Greenwood Press, 1991)
10. Charles T. Horngren and George Foster, *Cost Accounting, A Managerial Emphasis* (Prentice-Hall, Inc., 1991)
11. Peter B. B. Turney, from a forthcoming book, *Common Cents: The ABC Performance Breakthrough* (Cost Technology, 1991)

INDEX

ABC Project Worm, 106
ABC Technologies Inc, 66, 94, 146
Activity dictionary, 113, 126, 131, 153
Activity-based budgeting, 32, 49, 123
Activity-based Business Intelligence. *See* Business Intelligence
Analytic application, 10, 57
    ABC/M, 44
    Best-of-breed, 58–60
    Eliminate the essential, 43
    Markets, 95
    View, 57
Analytic systems. *See* Analytic application

Balanced scorecard, 12, 53, 140
Business intelligence, 54, 56, 70
    Activity-based, 64, 71
    Architecture of, 62
    Ledger-based, 69
    Software systems, 60
    Transaction-based, 68
Business process reengineering. *See also* Re-engineering

CAM-I, 14, 2
    Cross, 19, 153
    Focused studies in ABM, 32
    Hidden beauty of, 21
    Methodology, 35, 134
    Modeling method of, 134
    Standards in, 49
    Use of assignment paths, 49
    Way of, 49
Case studies
    Grupo Càsa Autrey, 133, 181–184
    Learning from others, 134
    Providence Portland Medical Center, 166–175
    U.S. Airways, 176–180
    Willard foods, 159–165

Change fatigue, 80
Correlated drivers, 121, 32

Data mining, 32
Data warehousing, 32
    Markets of, 95
    Technology, 60–62
Drucker, Peter, 23
    Father of modern management, 6
    Wealth creation, 7–8

ECR, 28–9, 158
Efficient Consumer Response. *See also* ECR
EDR. *See also* Electronic Data Replenishment
Electronic Data Replenishment
Enterprise-wide
    Activity-management, 34
    Deployment, 32, 42
    Implementation, 43
    Production systems, 116
    What is, 60
Electronic Resource Planning (ERP)
    Analytic systems, 10, 152
    Connections with, 32
    Differences between, 93
    Firms, 32
    Operational, 58
    Post-, 59
    Solution with, 59
    Systems, 43–44, 57, 59, 60
    Transaction processing with, 58
    Vendors, 57

Flowcharting, 140

GAAP, 12–13, 30
Gantt Chart, 127

Hidden costs, 143

INDEX

Information behavior, 89
Integrated ABC/M
   Analytic ABC/M, 96
   Modules, 59

KPI. *See also* Strategy
   Key activity sets, 148
   Performance measures program, 91
   Strategic alignment with, 147
   Strategy to, 197

Lean models, 118

Macro and micro activities, 119
Model certification and verification, 154–155
Model consolidation, 154
Moore, Gordon, 6
Multidimensional analysis
   Analysis of, 66
   Manager, 56
   Questions using, 76
   Views of, 51, 137

NAP, 15–19

On-line Analytical Processing (OLAP)
   Technology, 32
   Invention of, 62–64
   Applications and servers, 95–99
   Navigation engines, 123
On-line transaction processing (OLTP), 62–64
Organizations center of gravity, 83
OROS, 131
   ABCPlus, 163
   Calculation with, 162
   Model building using, 178
Output measure, 35

Paradox map, 83
Performance management
   Politics of, 30–31
Performance measures, 31, 53, 64
   Tags, 121
   Tracking with, 53
Phases
   Understanding the, 34
   Learning, 36
   Four distinct, 36
   Trigger, 36–38
   Education, 38–39
   Pilot, 39–40
   Enterprise, 41–44
   Objective setting, 106, 107–109
   Data gathering, 106, 109–111
   ABC modeling, 106, 112–122
   Integrating, 106, 111–112
   Reporting, 107
   Empirical data replenishment, 107, 122–126
   Forecasting and budgeting, 107, 123–124
   Re-creating reports, 107
Porter, Michael, 146, 159
Process mapping, 32

Rapid Prototyping, 87
Re-engineering
   Father of, 24
   Performance measurement and, 57
   Process, 158

Sevenfold way, 71–73
Strategy
   Links to, 7
   ABC/M to, 8, 140, 150
   Underpinning of, 9
   Datawarehouse as, 62
   Link models to, 114
   Activity not aligned to, 146
Subsystems
   Data collection and input, 47–49
   Modeling and analysis, 49–52
   Reporting and deployment, 52–53
   Predictive and planning, 53
   Infrastructure, 54
   Analytic and operational, 57
Surveys
   On-line, 32, 110
   Web-based input of, 49
   Tools, 50
   Paper, 110
   Electronic, 122

Task relevant leadership, 77–78
Task relevant readiness, 79
Tools inventory, 126
TQM, 79, 82

Wal-Mart, 5–6
Web
   HTML reports, 123

CUSTOMER NOTE: IF THIS BOOK IS ACCOMPANIED BY SOFTWARE, PLEASE READ THE FOLLOWING BEFORE OPENING THE PACKAGE.

The enclosed disk contains files to help you utilize the models described in the accompanying book. By opening the seal, you are agreeing to be bound by the following agreement:

This software product is copyrighted, and all rights are reserved by ABC Technologies, Inc. You are licensed to use this disk on a single computer. If the software is to be used on more than one computer in a company or educational institution, additional copies of the diskette(s) may be made for each person in the company or educational institution. However, this license does not extend to the material in the accompanying book which may not be copied except with the written permission of John Wiley & Sons, Inc.

This software product is sold without warranty of any kind, either expressed or implied, including but not limited to the implied warranties of merchantability and fitness for a particular purpose. In developing this software, neither the authors nor the publisher are engaged in rendering legal, accounting, or other professional services. If legal advice or other expert assistance is required, the services of a competent professional should be sought. Neither the publisher nor the authors or owner of the copyright assume any responsibility for errors, omissions, or damages, including without limitation, incidental, special, or consequential damages (including lost profits) which may result from use of the information in the book or on the diskettes. (Some states do not allow the exclusion of implied warranties, so the exclusion may not apply to you.)